© Courtesy of the author

About the Author

LAURA LEE is the author of eleven books, including *The Pocket Encyclopedia of Aggravations* and *100 Most Dangerous Things You Can Do in Everyday Life and What You Can Do About Them*. She brings to her writing a unique background, including stints as a morning radio show DJ, improvisational comedian, and a professional mime. She lives in Michigan.

Blame It on the Rain

Also by Laura Lee

100 Most Dangerous Things
You Can Do in Everyday Life and
What You Can Do About Them

The Pocket Encyclopedia of Aggravations

The Elvis Impersonation Kit

The Name's Familiar

The Name's Familiar II

Arlo, Alice & Anglicans

Bad Predictions

Invited to Sound

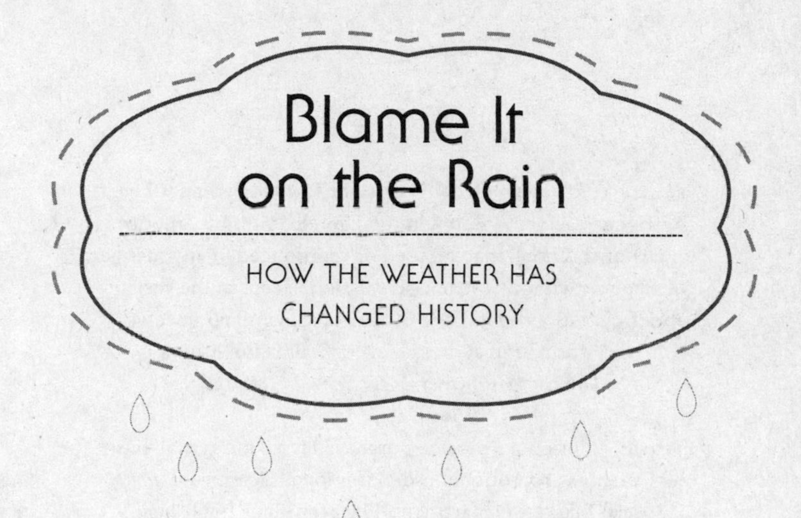

Blame It on the Rain

HOW THE WEATHER HAS CHANGED HISTORY

Laura Lee

HARPER

NEW YORK • LONDON • TORONTO • SYDNEY

HARPER

For information address HarperCollins Publishers,
10 East 53rd Street, New York, NY 10022.

HarperCollins books may be purchased for educational, business, or sales promotional use. For information please write: Special Markets Department, HarperCollins Publishers,
10 East 53rd Street, New York, NY 10022.

FIRST EDITION

Designed by Nicola Ferguson

Library of Congress Cataloging-in-Publication Data
Lee, Laura.
Blame it on the rain : how the weather has changed history / Laura Lee.
p. cm.
ISBN-10: 0-06-083982-1
ISBN-13: 978-0-06-083982-6
1. Weather—History. 2. Weather—Social aspects.
3. Climate and civilization. I. Title.
QC981.L43 2006
551.609—dc22 2005056392

06 07 08 09 10 ❖/RRD 10 9 8 7 6 5 4 3 2 1

As this book was being prepared for printing, Hurricane Katrina devastated the Southern United States—a sobering reminder that for all of our technology and scientific knowledge, we are still inhabitants of a planet governed by impersonal natural systems that are larger than our attempts to control them.

At the time of this writing (September 4, 2005), it is impossible to know what long-term effects the tragedy will have on United States or world history. But whether or not it is remembered as a historical turning point on a grand scale, it is without a doubt a defining moment in the personal histories of thousands.

This book is dedicated to all those throughout history whose individual histories were altered by storms and other forces of nature.

Contents

Introduction, 1
Humans on the Brink of Extinction, 7
Noah's Flood, 11
The Population of Australia, 15
Sea Breezes Save Western Culture, 18
Tempestas Cooritur: The Battle of Teutoburg Forest, 25
Why the Sun Never Sets on the British Isles, 30
The First Kamikaze, 35
Loss of the True Cross, 40
Greenland's Vikings: Victims of Climate Change and Cultural Stubbornness, 45
The Treaty That Fell from the Sky, 52
My Pope's Better Than Your Pope: Lightning and the Great Schism, 56
The Mud That Made England, 61

The Fog of War, 67

Lost Siberians, 73

Which Witch Did This?, 79

A Protestant Wind Destroys the Spanish Armada, 84

Thanks to the Wind the Lost Colony Remains America's Greatest Mystery, 88

Gee, It's Cold in Russia, Part I: Charles XII Invades Russia, 92

The Secret of the Stradivarius, 97

Another Protestant Wind Blows a New King to England's Throne, 100

Ben Franklin and That Kite, 105

Through Many Dangers, Toils, and Snares, 110

Washington and the Weather, 117

Hail to the French Revolution, 124

Rain Ruins Robespierre, 129

The United Irishmen, the French, and the Rain, 135

A Slave Revolt Washed Away, 139

Gee, It's Cold in Russia, Part II: Napoleon Invades Russia, 145

Does That Star-Spangled Banner Yet Wave?, 152

Tecumseh Is Lost in the Fog, 159

The Water of Waterloo, 166

Gee, It's Cold in Russia, Part III: A Senseless War Extended by the Weather, 170

The Guy with the Sideburns Gets Stuck in the Mud, 178

The Storm That Saved Civil War Prisoners, 182

What Is That Guy in "The Scream" Afraid of?, 186

A Gust of Wind and Aviation Obscurity, *190*

El Niño and Dashed Polar Dreams, *194*

Cold Shaving and Jacob Schick, *199*

The Failure of Forecasting and the Death of Lord Kitchener, *203*

Rain Clouds Put an End to the Age of the Airship, *209*

Gee, It's Cold in Finland: The Winter War, *214*

Gee, It's Cold in Russia, Part IV: Hitler Invades Russia, *219*

D-Day, *230*

Blooming with Atoms: A New Cloud Formation, the Mushroom, *237*

Sunshine over Hiroshima, *243*

Misreading the Monsoons, *249*

Dewey Defeats Truman, *254*

Canadian Chill Saves a National Park from Nuclear Contamination, *257*

Heat and the Powder Keg, *261*

Making Monsoons, *267*

Lucy and Her Friends, *274*

Operation Thwarted by Desert Storm, *277*

Meteorology and Rocketry, *283*

Blood Rain and World War III, *289*

Nature Does Not Carry a Passport, *293*

Bibliography, *297*

Blame It on the Rain

Introduction

> Russia has two generals whom she can trust—
> General January and General February.
>
> —Czar Nicholas I

For years Russia has been home to a secret weapon, more powerful than the nuclear warheads that made it a world superpower. When all hope seems lost and the casualties of war are mounting, there is a cold war the Russians almost always win: the battle with the elements. Russians are proud of their hideous weather, and they'll often describe a –40°C (–40°F) day with the same pride a fisherman uses to describe an especially large marlin he caught. Russia is home to the coldest city on Earth, Yakutsk. Interestingly, an English researcher has found that the British, not the Russians, are more likely to die of the effects of exposure. Why? Because no Siberian has ever been shocked to discover it was cold outside. Siberia has cold like Bill Gates has money. In January, a typical Yakutskian will not venture out in less than 4.26 layers of clothes. (That's an average arrived at by a scientist. They don't cut their clothing in quarters.) A Londoner, meanwhile, goes out with little more than a jacket. This understanding of the elements has served the Russian people very well. In fact, were it not for this skill, Russians might be speaking French today.

In 1812 Napoleon assembled the largest army Europe had ever seen—more than six hundred thousand strong. His plan was to march boldly into Russia. He was not at all worried that winter was approaching. Napoleon's confidence appeared well founded when his soldiers captured Moscow. They pillaged the city and stole jewels and furs as war prizes to present to their wives back home.

Then the one thing that Napoleon had failed to consider became abundantly clear: Russia can get very, very cold. As Napoleon's army marched away from the ruined city with their spoils, temperatures fell to –40°C. The soldiers fell to frostbite and starvation. In one twenty-four-hour period, fifty thousand horses died from the cold. The men wrapped up in their wives' war prizes, but to no avail. Of the six hundred thousand men who marched into Russia, only one hundred and fifty thousand would limp home. It was the beginning of the end for Napoleon's empire, and heralded the emergence of Russia as a power in Europe.

Napoleon would not be the last to underestimate the power of climate as a military weapon. Adolf Hitler, apparently not much of a student of history, decided to repeat Napoleon's attack on Moscow, including the part about freezing to death. In September 1941, Operation Typhoon (one of many military operations named for extreme weather) swept into the Soviet Union. The German army was so confident it would win against Stalin's troops that several units brought dress uniforms along for the victory march in Red Square. What they didn't bring along, however, was winter clothing. In early December the mercury dropped to –35°C (–31°F), and heavy snow began to fall. German soldiers, dressed in summer clothing, tried without success to dig for shelter in the frozen ground. Their war machines were not designed to function in the frosty conditions—supply vehicles broke down, rail tracks shattered, planes could not fly, machine guns iced over. Fresh Siberian soldiers, however, had lived with such conditions all their lives. Dressed in layers with felt boots and warm coats, and with tanks designed to operate in ice and snow, they attacked the faltering Germans. Then just as the cold began to abate, the Germans encountered another Russian weather phenomenon, the *rasputitsa*.

Rasputitsa is the time of year when snow melts and the roads become an impassable quagmire. You see, in Russia the temperature can rise and fall with amazing speed. One hour it is below freezing, the next it has climbed above, only to plummet again. This combination creates puddles of mud, potholes that are hard to see because they are filled with slush and slick ice covered in mounds of mushy earth or snow. It should have been enough to make even Hitler rethink the value of this conquest, but he realized his mistake too late. His meteorologically assisted defeats in the Soviet Union, both outside Moscow and in Stalingrad, were turning points in the war.

Talking about the weather has come to be synonymous with meaningless chatter, but don't be fooled. The winds have a power to shape nations and cultures. Overcast skies impact our outlook and our expectations. The rains have the power to alter our moods, politics, hospital admissions, and even to change history.

Talking about the weather wasn't always seen as trivial. For our ancestors, being in tune with nature was not a matter of choice but of necessity. The weather determined when to sail, when to sow, when to forage, and when to venture out into the wilderness. It was a matter of life and death. Great social upheavals often followed periods of extreme weather. You will read about governments that toppled due to weather-induced plagues, riots set off by extreme heat, and religious panics brought on by lightning strikes.

Today, with the exception of Weather Channel employees, we tend to leave the forecasting to others. We move from one climate-controlled environment to another and pay little attention to the aspiring stand-up comedian who dispenses puns while pointing at the electronic map on the local news station. By leaving concern over the weather to the professionals, we may fail to realize how it continues to impact our future in subtle and not-so-subtle ways.

One of the first to examine the impact of weather on society was Baron de Montesquieu. In his 1748 book *L'Esprit des lois* he presented twenty years of comparisons between human characteristics and governments and the climates that acted upon them. He concluded that inhabitants of cold countries had no real sensitivity to subtle tastes or to wasteful emotion, while those who lived in warmer climes were more emotionally open and volatile. Montesquieu's observations may have been broad stereotypes, but sociologists, archaeologists, psychiatrists, and physicians continue to explore the links between the weather and our societies and how the weather impacts our everyday lives.

Temperatures impact nations' traditional costume—grass skirts would make no more sense in Finland than fur hats would make in Hawaii, and the average English businessman would not be known for his bowler and umbrella if the country's rainfall was more similar to that of Burundi. Gods and religious fables are used to explain the tides, floods, and winds. Most of the world's major holidays are tied to seasonal changes and harvests. Poetry, prose, music, and visual art owe great allegiance to the beauty and fury of the weather. Our language is full of expressions like "under the weather" and "on cloud nine." ("Cloud nine" is an old meteorological term for clouds they now call noctilucents. These are some of the highest clouds in the sky.)

The foods we eat are determined by our climates, and when the weather fails to produce a bumper crop, whole communities migrate. Businesses fail or thrive based on adequate sunlight. Weather was also one reason why America's film industry is located in Hollywood and not in its previous hub, Fort Lee, New Jersey.

Leaders rise and fall on the whims of the weather. Not only because of victories and defeats in battle but also by the impact of storms on the ballot box. Political strategists spend a fair amount of time speculating how the rain will impact voter turnout, and who will benefit.

While most of us are content to find better ways to adapt to the weather, some strategists look for ways to take control of it. A recently declassified Department of Defense Report, "Weather as a Force Multiplier: Owning the Weather in 2025" describes not only new forecasting techniques but also storm-creation technologies. Attempts at weather control have, in fact, been with us for generations. On a small scale there is air-conditioning, which allowed cities to prosper in the hottest climates. We have snow machines to take some of the risk out of the ski business. And don't think military units have been waiting until 2025 to tamper with the skies. As early as 1957 a report of President Eisenhower's advisory committee noted that weather control could become a "more important weapon than the atom bomb." In 1966 the United States got a chance to try it out. Project Popeye dispensed a rain-making agent into the clouds of Vietnam and extended the monsoon season in order to increase the mud on the Ho Chi Minh trail—the main enemy supply route to South Vietnam.

Even when we do not intend to change the weather, human activity somehow always does. Patterns of land use—overgrazing grasslands, plowing prairies, and irrigating fields—significantly modify local weather. Miles of pavement and heat rising from cities create their own weather systems. Jets put man-made clouds in the sky. Following September 11, 2001, scientists noted that in the absence of the Twin Towers, lightning patterns changed over Manhattan.

We are, it appears, not merely at the mercy of the elements. We are in a marriage with the elements. Societies cannot exist without impacting the air around us, and what we do to the sky is destined to eventually rain down on us. Human society is shaped by the weather, impacts the weather, and then must adapt to these newly created patterns; each one of us is only a small part in this complex and interdependent system.

The following chapters will explore some of the moments when climate, weather patterns, and storms influenced the course of history. Storms that were devastating to a particular region but did not have larger historic implications will not be covered. Nor will natural disasters that are geological, not meteorological, in nature, like earthquakes, tsunamis, and erupting volcanoes (unless those events impacted the weather).

It is not my intention to suggest that weather was the only cause of the events in this book. That would be an oversimplification. When various elements come together, an ill-timed typhoon can tip the scales in favor of one side in a war, and that can have far-ranging consequences.

Humans on the Brink of Extinction

We humans are an egocentric lot. We tend to look at the world and all of its history and prehistory as leading to that great moment when humans would reign supreme, the ultimate goal of creation.

As the humorist Douglas Adams once observed: "This is rather as if you imagine a puddle waking up one morning and thinking, 'This is an interesting world I find myself in—an interesting *hole* I find myself in—fits me rather neatly, doesn't it? . . . Must have been made to have me in it!'"

The truth, as hard as it may be for us to accept, is that our preeminence on the planet was not preordained. Our human ancestors could well have gone the way of the dinosaurs, and in fact, they nearly did.

Since the beginning of life on earth, there have been periods of mass extinction, bringing down a once-dominant species and making way for a new life-form to have its moment in the sun. The last such mass extinction, sixty-five million years ago, destroyed the dinosaurs and allowed the mammals to take over. Geologists and archaeologists spend their entire careers exploring the causes of mass extinctions. The reigning theory at present is that many can be blamed on extreme weather conditions created by natural disasters.

The dinosaurs were most likely the victims of a wayward meteor that struck the Yucatán Peninsula of Mexico and bored a twenty-five-mile crater into its surface. The explosion was the equivalent of the detonation of one hundred million hydrogen bombs—the heat vaporized seawater and saturated the atmosphere. The superheated air swept outward. Dust from the blast traveled so far that it blanketed what is now Kansas and flew into the atmosphere, where it encircled the globe and blotted out the sun, cooling the planet. Plants could not photosynthesize, so they died, and the creatures that depended on them soon followed. It is called the Great Dying. Nearly 90 percent of all life that then existed was wiped out forever. Fortunately for us, one of the survivors was the cynodont, the ancestor of modern mammals.

Our own very similar brush with extinction came about seventy thousand years ago. DNA studies point to a population crisis, sometimes called a population bottleneck. Scientists sought to understand why there was so little genetic variation among humans. There is more genetic variation in a single group of chimpanzees or a clan of gorillas than there is in the entire six-billion-member human population. This points to a time when there were only a few procreating females around. One study suggests the number dropped to as few as five hundred; it would take another twenty thousand years for the human population to fully recover and regain its previous numbers.

The cause of the bottleneck was one of the largest volcanic eruptions in four hundred and fifty million years. The explosion of the Toba volcano on what is now the Indonesian island of Sumatra produced a crater that measured 100 km (60 mi) across and a plume that was at least 30 km (19 mi) high, scattering rock and ash as far away as Greenland. It tossed about 2,800 cu km (684 cu mi) of molten rock into the atmosphere. That's enough to build more than a million Great Pyramids of Egypt. The blanket of ash blocked out the sun, and global temperatures dropped by up to 12°C (22°F). The volcanic winter lasted for six years. The increased snow cover that accumulated in this period further reflected the sun's rays, preventing the ground from absorbing heat, which made the world still colder. It was the beginning of a thousand-year ice age.

Some researchers speculate that an ice age that was already in progress was the cause, not the effect, of the eruption. The ice age may have lowered sea levels, relieving pressure on the volcano and allowing it to blow, like a cork being removed from a bottle of champagne. The effects of the eruption then sped the glaciation in progress, providing the trigger that changed the climate system from warm to cold, and as glaciers formed, the sea level dropped further. The exposed soils were carried away by the wind. Dust storms raged for days, killing plants and animals.

Homo sapiens was very close to going the way of the Neanderthal and other extinct human species, but a few hardy individuals survived in isolated pockets in Africa, Europe, and Asia. As a result, our population has only a small sample of the genetic diversity we once had.

Could such an eruption happen again? Absolutely. Not only could it happen, it is almost inevitable that it will. The most likely site of the next supereruption is Yellowstone National Park. The geysers, hot springs, and mountains that bring tourists to the area are caused by a large underground magma chamber that extends about 20 km (12.5 mi) across and 2,900 km (1,802 mi) down—nearly halfway to the center of the Earth. Yellowstone has already exploded three times. It blows every six hundred thousand years or so. The last six hundred thousand-year mark was reached four hundred thousand years ago. Yellowstone is only one of forty supervolcano sites, but most are extinct, and none is as close to heavily populated areas.

When the Yellowstone volcano blows, scientists say it will unleash a force that is larger than the entire planet's nuclear arsenal. The blast would be heard as far away as England. About one hundred thousand people would die immediately. Toxic gases and ash would be thrown into the atmosphere, and it would fall across the entire western United States within hours. It would continue to spread across the globe on the winds, creating a volcanic winter. It could happen next week, or two hundred thousand years from now.

Noah's Flood

Of all the meteorological events that supposedly changed history, the greatest would have to be the flood that wiped out all of creation except for Noah, his family, and all but two of every type of animal. But did the biblical flood actually happen?

William Ryan and Walter Pitman, the authors of the book *Noah's Flood,* believe that there was indeed a great flood that inspired the Noah story and many of the flood legends recorded throughout the world. More than two hundred similar flood myths have been recorded, in the cultures of the Greeks, Egyptians, and Babylonians. The Sumerian epic *Gilgamesh,* inscribed on clay tablets around 2000 BC, says that Gilgamesh was warned by a god to build a great ship to protect all the living things on earth from a forthcoming flood.

All of these stories may well have a common source, a deluge so great that it seemed as if the world was immersed: a sudden disaster that survived in folklore and was passed along from generation to generation, becoming even more dramatic with each retelling. The researchers believe they have found exactly such an event. It happened about seventy-five thousand years ago in the land surrounding the Black Sea.

Fossil evidence shows that until that time the Black Sea was a modest freshwater lake fed by meltwater from ice age glaciers. Along its shore were numerous settlements, and a small ridge divided this lake from the Sea of Marmara. All at once, however, the fossils in the Black Sea shifted from freshwater mollusks to saltwater mollusks. Research by explorer Robert Ballard, using sonar and a remote-controlled underwater camera, shows an ancient shoreline 167 m (550 ft) from the current coast of Turkey. Stone tools, mud-and-wattle walls, and shards of pottery were buried beneath the sea for millennia.

Some scholars have suggested that the shores of this body of water—we'll it call the "Black Lake"—may have been the cradle of civilization. The land would have been more fertile than Mesopotamia in the arid Middle East, and artifacts, language patterns, and ethnic relationships could be explained by the movement of people from the Black Sea region following a flood. This theory remains controversial. Still, there is little doubt that there were settlements around the Black Lake that were washed away by a great flood.

What happened? During the last ice age, glaciers extended down from the North Pole as far as Chicago and New York City. Much of the world's water was locked up in solid form, so the oceans were about four hundred feet lower than they are today. As the Ice Age drew to a close and the glaciers started to melt, the waters of the Mediterranean rose. They began to flow into the Sea of Marmara. As they rose, the pressure on the ridge dividing it from the lake became greater and greater. Finally, the natural dam was unable to hold.

It released a torrent of saltwater with the force of two hundred thousand Niagara Falls. The roar of the water could be heard at least one hundred miles away. Each day another 42 km^3 (10 cu mi) of seawater poured in. As it did, it inundated villages in what are now Turkey, Bulgaria, Moldova, Ukraine, Russia, and Georgia. The influx of water and its evaporation caused torrential rain across the region. Villagers ran for high ground, but the water kept coming. Every day it would advance another half a mile (.8 km). When it finally leveled off, the lake was a saltwater sea, 150 m (500 ft) higher than it had been before.

Was this event the origin of the story of Noah's flood? It is impossible to know. If it was, it would have been handed down as oral history for one hundred thousand generations, which could easily account for some of the variation in the details—you know how different your family stories get after only one generation.

Researchers Valentina Yanko-Hombach at the Avalon Institute of Applied Science in Winnipeg, Manitoba, and Andrey Tchepalyga of the Institute of Geography in Moscow believe that the flood would have been much less dramatic than Ryan and Pitman envisioned. Their analysis of sediment and seismic data from the Black Sea shows that water from the Caspian Sea flooded into the area about fourteen thousand years ago and that the Black Sea began to rise gradually over the course of one thousand years, driving the inhabitants out of the basin. After the Caspian overflow stopped, the level of the Black Sea fell, to be flooded by seawater a few years later. Yanko-Hombach believes this second flood raised the sea only forty meters and did so more gradually than the American team suggested. The various theories are still the subject of archaeological speculation.

As to the theological question, Rabbi Amy Wallk Katz told the *Kansas City Star,* "Archaeological truth is determined by the scientific method. . . . [H]istorical truth is determined by how a community recalls its past. . . . The biblical stories are significant because they teach us ethical and moral lessons of history, not because they reveal the facts of history."

Rev. Holly McKissick added, "The truth of the flood is not that animals went on two by two. The truth is this: God holds out a world of hope."

The Population of Australia

No one can say with absolute certainty how or why human beings first landed on the Australian continent. The people of that time were not thoughtful enough to write this kind of thing down. The origins of human life in any region of the Earth are nothing but educated guesswork—highly educated perhaps, but guesswork nonetheless. But one way or another, it is likely that weather played a significant role in taking people Down Under for the first time, and the journey is an amazing tale.

Our story begins about fifty thousand years ago. Australia was already a large island, disconnected from the Asian continent by at least 15–20 km. It was populated by many unique plant and animal species, including some large mammals that are now extinct, but there were no apelike creatures from which human beings could have descended. The first human beings were foreigners. They could only have arrived by sea.

It is only fairly recently that archaeologists have discovered that people have been in Australia for so long. A century ago, scholars assumed that the Aborigines had lived there for only four hundred years or so. In the 1960s scientists pushed the time frame back eight thousand years. Then in 1969 a geologist named Jim Blower, from the Australian National University in Canberra, was cxploring a long-dried lake bed called Mungo when he came across the skeleton of a woman sticking out of a sandbank. She was excavated and carbon-dated. Much to everyone's surprise, it turns out she was more than twenty-three thousand years old.

This means that humans must have traveled from Asia to the southern continent at a time before scholars believe humans had developed edge-ground stone tools like axes. Without these kinds of tools, the boats on which the early Aborigines traveled could not have been fancy. Since no examples of fifty-thousand-year-old boats exist, archaeologists have filled in the blanks with their imaginations. Most likely the voyagers journeyed on rafts made of lengths of bamboo lashed together with palm leaves (think of a large, flat wicker chair).

Now the question becomes: Why were they out on the sea to begin with? From where they stood, in southern Asia, these people could not have seen the distant shore of Australia. They could not have been certain that there was land out there. One theory holds that the journey came about more by accident than design.

It may have gone like this: our early Australian ancestor—we'll call him Joe—was out fishing on his floating platform. Ominous clouds were forming overhead, but Joe wanted to get just one more fish. Joe's wife waded out to the raft to tell him to come back. She climbed onboard just as a major storm, perhaps a summer monsoon, bore down upon them. Joe was unable to steer his raft in the stormy seas, and the wind blew the couple farther and farther out to sea. They drifted helplessly at sea for days until they finally washed up on a beach in northern Australia. With nothing else to do, they begin populating a continent.

The problem is that two people are probably not quite enough to create a society. For adequate breeding stock, Joseph Birdsell, an

American scholar, calculated that around twenty-five would be necessary. Maybe Joe was blown off course with his entire family in tow, but it seems highly unlikely.

Some adherents of the storm theory believe that our protagonist was fishing alone when storm winds blew him off to a strange land. The fisherman immediately recognized the opportunity, navigated his way back home, and led a group of colonists to this new shore. Another theory holds that it was a series of rafts and a series of unlucky Asians who individually misjudged the seas and found themselves the accidental colonists of the Land of the Lost.

Philip Clarke, author of *Where the Ancestors Walked: Australia as an Aboriginal Landscape*, argues that the population of Australia was not an accident at all. While there may have been no storm, these early explorers could only have known of the existence of Australia by observing the patterns of the weather.

Even though they couldn't see Australia over the horizon, the explorers might have observed "land clouds," which form above mountains when moist air chills and rises over a peak. The clouds would remain stationary during the day—they would seem to cling to the mountaintops. The explorers may have seen them and recognized them as signs of land. If the clouds didn't give Australia away, lightning could have. During a storm, lightning might have struck the part of northwestern Australia that is now undersea. If the lightning hit the brush and caused a fire, the clouds of smoke might have been visible from the costal mountains of Timor.

"From a single family group who made a successful crossing," wrote Clarke, "a whole continent could have been colonized in a matter of a few thousand years."

Long before the other peoples of the world had boats, the Aborigines were guided, through design or chance, by the weather. They traveled into the unknown and created a new society—one of the oldest cultures and languages that still exist on earth.

Sea Breezes Save Western Culture

The survival of Greek culture, and consequently of Western culture itself, hung in the balance during the Greco-Persian Wars. The Persian Empire, at the peak of its strength, was poised to overrun mainland Greece itself. The Greek naval commander Thermistocles was able to literally turn the tides of war at the battle of Salamis in 480 BC by utilizing his knowledge of the winds.

Knowledge of wind patterns was vital to all early seafaring cultures, but it seemed to occupy a special place in the Greek psyche. People of numerous cultures, including the Trojans, Spartans, Phoenicians, Carthaginians, and Etruscans, lived around the Mediterranean and they fought for control of trade routes. With their naval vessels at the mercy of sea currents, they all kept a close eye on the direction of the breeze, but according to meteorologist J. Neumann, the Greeks were by far the most vigilant. Neumann has found that references to the wind in ancient Greek literature are far more numerous than those found in the canon of the surrounding empires.

The earliest known Greek literature, the Homeric epics, dating back to 800 BC, frequently make reference to the winds, as in this passage:

> Thus all day long the young men worshipped the god with song, hymning him and chaunting the joyous paean, and the god took pleasure in their voices; but when the sun went down, and it came on dark, they laid themselves down to sleep by the stern cables of the ship, and when the child of morning, rosy-fingered Dawn, appeared they again set sail for the host of the Achaeans. Apollo sent them a fair wind, so they raised their mast and hoisted their white sails aloft. As the sail bellied with the wind the ship flew through the deep blue water, and the foam hissed against her bows as she sped onward. When they reached the wide-stretching host of the Achaeans, they drew the vessel ashore, high and dry upon the sands, set her strong props beneath her, and went their ways to their own tents and ships.

Aristotle's *Meteorologica,* the first book to attempt to categorize weather phenomena in a systematic way, gives the various winds their own names. His student, Theophrastus (374–287 BC), continued with his work in the treatise *On the Winds*. Yet with all of their knowledge of the winds and waves, the Athenians of the fifth century BC lacked a true navy—they were forced to lease ships from the Corinthians. While Persia achieved its greatest victories on the land, they also amassed a naval fleet that was one of the largest and most feared in the world.

The foundations of the battle of Salamis go back to the sixth century BC, when the Persian kings extended their rule from the Indus River Valley to the Aegean Sea and eventually conquered the Greek city-states along the Anatolian coast. In 500 BC the people of these cities rebelled against Persia; mainland Greece sent reinforcements, but the revolt was unsuccessful. The Persian king, Darius, used the battle as an excuse to invade Greece, but the weather intervened. A storm destroyed most of his fleet, and he was forced to turn back.

In 490 BC, the Persians invaded again; this time a force of twenty-five thousand men landed unopposed on the plain of Marathon. The greatly outnumbered Athenians prayed to their gods and promised to sacrifice a goat to Artemis for every Persian killed in the engagement. The gods seem to have heard the Greek prayers, and, against all odds, they won a decisive victory. The achievement was a cause for great celebration and the runner Pheidippides ran to the city of Athens, to spread the good news. (It would not be until 1896, when the first modern Olympic Games were held in Athens that the word "marathon," meaning a race, would enter our vocabularies. The Olympics included a long-distance race to commemorate this run.) The dead of the battle were buried on the field itself under a mound of earth that is still visible today. After the battle, there were not enough goats in Athens to satisfy the Athenians' debt to Artemis. They offered up five hundred and repeated the sacrifice each year for almost a century.

After Marathon, the Athenians forgot all about the Persian threat. In fact, it would be nearly another decade before Themistocles would convince the Athenians to build a navy. Even when they did undertake this task in 483 BC it was to protect themselves against the neighboring Greek city-state Aegina, not the Persian Empire. The Persians also had other things on their minds. In June of 486 BC the Egyptians and Palestinians revolted against Persia. Until those matters were taken care of, Greece would have to wait.

Thus it fell to Darius's successor Xerxes to try to finish the job and claim Athens for Persia. Xerxes was taking no chances—he assembled the greatest armada the world had ever seen. Historians debate just how large it was, but a safe guess is about one thousand ships ferrying two hundred and fifty thousand men. (To put this into perspective, the population of the planet at that time is estimated to have been about 162 million, roughly the population of modern-day Pakistan.) Along with the Persians were squadrons of Ionians, Cilicians, and Phoenicians. The Phoenicians were no slouches themselves when it came to sailing. They dominated the waters of the Mediterranean from 1200 to 900 BC with knowledge of such things as how to navigate using the North Star and may well have been the first Western people to sail around the southern tip of Africa.

The Persians made little secret of their attack plans. Given his overwhelming power, Xerxes believed intimidation was a more powerful weapon than secrecy. He sent heralds in advance of his troops to demand water and order the communities to prepare meals for the king and his army. The Athenians knew Xerxes' plan and his movements. They knew how many people the communities were expected to feed and thus, how many soldiers. Xerxes' military might was like a storm the Athenians could clearly see gathering over the horizon.

In September the Persians burned an evacuated Athens and were poised to conquer the Greek navy. If the Greek fleet of about 380 ships were to face the massive Persian fleet in the open sea, they would have no chance at all. So Themistocles devised a plan to lure the invaders into a channel between the island of Salamis and Piraeus at just the moment when a brisk wind would blow from the open sea.

First, he sent a messenger to "accidentally" reveal that the Greeks were planning to retreat from Salamis under cover of darkness. Xerxes took the bait. He reasoned that if he could trap the Greeks as they tried to escape, he could do away with their whole navy at once and avoid a protracted series of battles. He ordered his ships to block the narrow Megara Channel on the west side of Salamis, where he believed Themistocles' ships would try to flee.

About two hours before dawn, Themistocles assembled his men. The sixty thousand were vastly outnumbered, and the chance of success was slim. The commander addressed them, as the historian Plutarch records:

"The white heralds of dawn have rekindled the dawn with diamond rays," he said. "Valiant sons of Greece: Go forth, and deliver thy kindred. Thy ancestors' tombs and the sacred temples your fathers did build! But one more combat to end the ravenous war. Now, all is at stake. Sons of Greece: Go!"

Shortly before dawn, the Greeks sailed. But instead of retreating to the south, they sailed around the north end of Salamis. Themistocles knew that about two hours after sunrise the Etesian breeze would make it hard for the Persian fleet, weary from battles and the journey across the sea, to navigate.

As the Persians chased the Greek triremes into the narrow strait, the brisk wind roared across the sea, right on schedule. The sea began to swell, rocking the top-heavy Persian vessels. The narrow, low-slung Athenian vessels had an easier time of it. They held their position in the channel at its narrowest point, where the channel was only 1,371.6 m across. It was here that Themistocles wanted to do battle, and he drew Xerxes' men straight to the spot.

They held this position as the tremendous Persian fleet bore down upon them. A thousand ships or more choked the waterway. The Greeks sailed straight for the enemy and rammed their ships. By midmorning, the southern breeze was blowing straight up the channel, and the Persians, with the wind to their backs, found it hard to maintain a steady course. The Greeks, in the lee of the Cynosura Point, were meeting the wind head-on, which gave them greater stability. By noon, the Persians had lost their formation entirely. Their great numbers left little room for maneuvering in the narrow strait. As the ships in the front were disabled, those in the rear, unaware of the danger ahead, kept sailing. The Persians were being rammed by both sides, in front from the Greeks, in the rear by their own ships.

By midafternoon, the sea was nearly solid with broken hulls, oars, and sailors' bodies. More than 240 wrecks littered the sea, and floating on or nearby were the bodies of fifty thousand men.

"The shores of Salamis, and all the neighboring coasts are strewn with bodies miserably done to death," wrote the Greek playwright Aeschylus.

When it was all over, the Persians had lost one of their three squadrons, and bodies would wash ashore for days to come. While they still outnumbered the Greeks, they had lost all semblance of command and control, and the remaining ships sped toward home. They had burned Athens to the ground but had failed to defeat the Athenians.

The weather would deal yet another terrible blow to the retreating Persians. The stormy autumn season was fast approaching and after that, the winter. The Persians had to flee as quickly as possible to the waters of Asia Minor. In October the Persian army withdrew to Thessalia (a region of northern Greece, south of Macedonia, east of Epirus and bordering the Aegean Sea). From there they had but forty-five days to reach the Dardanelles (the Gallipoli Peninsula) before the winter weather would make travel difficult. Since they were moving at such a swift clip, supply trains could not keep up, and famine and illness caused by drinking polluted water ravaged the men. When they finally reached the Dardanelles, the parched, starving men gorged on food and water, but it was no help—the sudden change in their diet was enough to kill some of them. The most powerful of the contingents comprising Xerxes' armada deserted him after Salamis. Never again would Xerxes invade Greece with such force.

Had Xerxes been victorious at Salamis, he undoubtedly would have attacked the costal city-states one by one. They could not have organized a common resistance, and ancient Greece, with its classical mythology, philosophy, and concepts of democracy, might have been wiped away. Instead, for the Greek navy, the Battle of Salamis was a patriotic triumph which inspired Athens to create a naval empire. With their naval power crippled for a decade, the Persians would have to go on the defensive—first against Athenian naval actions in the eastern Mediterranean and later against Spartan and Macedonian invasions of Asia. As David Sacks noted in the *Encyclopedia of the Ancient Greek World,* "the process that would end with Alexander the Great's conquest of Persia (334–323 BC) began in the narrows at Salamis."

Tempestas Cooritur
The Battle of Teutoburg Forest

A storm is on the way.

At the height of its empire, Rome was the most super of superpowers. By 7 BC it had conquered all of the Iberian peninsula and established a chain of fortresses along the right and left banks of the Rhine. Rome controlled Austria south of the Danube, east Switzerland, and southern Germany. Its military power was unrivaled, and the conquest of the world seemed all but inevitable. In the first year AD Tiberius, who would later be emperor, quelled an uprising by the Germanic Cherusci and Lombard tribes, known to the Romans simply as "barbarians." Most of northern Germany fell under Roman rule.

Back in Rome, Emperor Augustus felt confident that all was well in the north, so he recalled Tiberius and sent him to quell a revolt in Pannonia, a central European province that included parts of modern Austria, Hungary, Slovenia, Croatia, and Serbia and Montenegro. In his place, Augustus sent Publius Quintilius Varus into Germany.

Varus was a wealthy man from an old patrician family and was married to a favored grandniece of Augustus. He had recently returned from the proconsulate of Syria, where he plundered the region for his personal gain, and in the words of the British historian Sir Edward Stephen Creasy, "habitually indulged in those violations of the sanctity of the domestic shrine, and those insults upon honor and modesty."

Varus, as the head of what we might today call an army of peacekeepers, no doubt expected this to be an easy assignment. The Cherusci, to all appearances, were friendly and loyal, content to be part of the great empire. He got right to work exacting tribute from those he governed and recreating with their wives and daughters.

Varus, however, had completely misread Arminius, the chief of the Cherusci. Arminius was one of many Germans recruited to fight for Rome. He had obtained Roman citizenship and equestrian rank. Yet as he flattered Varus, Arminius was gathering information, stirring up anti-Roman sentiment among the Germanic tribes, and preparing a surprise attack, the only type of battle that had any chance of success against an army of the world's military superpower. His plot would be aided in no small part by the weather.

Once Arminius had assembled his fighters, he sent word to Varus that his people needed protection from a revolt to the north. The legionaries would have to cross Cherusci territory to reach the site of the uprising. Varus, relatively unconcerned, packed up his camp and began to move. With him were the XVII, XVIII, and XIX legions; three units of cavalry; and six cohorts of auxiliaries: about twenty thousand men. They were not expecting to march into battle, so they traveled, not in formation, but in a relaxed march along with a wagon train of their possessions and their spouses and children. The whole assemblage was led by Cherusci scouts.

The Germans led the legions into a hilly, wooded region surrounded by steep mountains. The passageways through the rocks were narrow. What is more, recent storms had left the ground muddy and the paths flooded by overflowing streams. The supply carts were heavy and bogged down in the quagmire.

Arminius used this opportunity to attack the rear guard. The Germanic guides fled, leaving the Romans in unfamiliar territory. Throughout the day, the Cherusci attacked in hit-and-run fashion until Varus ordered the legions to stop and build fires around camp. The camp was not attacked that night. The next morning Varus ordered the legions to leave the unwieldy wagons behind; as the soldiers scrambled to get their possessions, Arminius gave the signal for the attack.

The fighting continued through day two and into day three, when a gray cloud appeared and signaled the beginning of the end of the Roman legions. A violent downpour with pounding rain, rolling thunder, and crackling lightning filled the sky, toppling trees, breaking up the ranks, and turning the bloody field into a swamp. The rain soaked the leather covering of the Romans' shields, making them too heavy to hold. They were pelted by hail and thrown by panicked horses.

Besides the tactical considerations, the storm had symbolic meaning for each side. To the Romans a storm was a sign of the gods' displeasure, a bad omen. A single stroke of lightning was enough to cancel all public assemblies. The phrase "I will watch the sky" became synonymous with political vetoes.

"Is it possible for us to doubt the prophetic value of lightning?" wrote the first-century historian and biographer Quintus, "When the statue of Summanus which stood on top of the temple Jupiter was struck by a thunderbolt and its head could not be found anywhere, the soothsayers declared that it had been hurled into the Tiber, and it was discovered in the very spot which they had pointed out." If one bolt of lighting was a bad omen, imagine how the Romans must have perceived a tempest in the middle of an unexpected battle.

To the Nordic people Thor was the god of weather and war, and Thor's thunder was a sign of divine assistance. The Cherusci now had proof that the gods were on their side, and they attacked the dispirited legions with new vigor.

Surrounded, exhausted, and ankle-deep in mud, the Romans could no longer retreat, even if they wished to. If the Romans expected mercy from the Germans, they were sadly misguided. The Cherusci killed their Roman occupiers with ferocity. They hacked Roman bodies to pieces and offered their corpses up in rites on altars to Thor. Some survivors were placed in wicker cages and burned alive. Varus, unable to bear the humiliation of the loss, fell on his own sword. His head was sent to Rome in a canvas sack. The number of Roman dead was estimated to be about twenty thousand.

It was a brutal, humiliating, and painful loss for the Romans. When Augustus received news of the loss, he cried "Varus! Give me back my legions!" Never again would Rome assign the numbers XVII, XVIII, or XIX to a legion.

Rome was not about to let Arminius have the last word. Germanicus Tiberius Caesar set out with a new contingent to prove that the Romans were still the mightiest military force in the land. Five years after the first battle, the Roman legions marched back into Teutoburger Forest. This time they were ready for combat, and they captured Arminius's wife, Thusnelda, and were the easy victors in battle. But it was all for show—Emperor Augustus and his successor, Tiberius, never again tried to conquer northern Germany and the German "barbarians." The Goths and the Vandals would eventually cause the downfall of the Roman Empire. German tribes, free from the yoke of Rome, would spread across Europe. The Cherusci merged with the Saxons of northwest Germany, who would eventually become the Anglo-Saxons who invaded and colonized Britain and became the English.

Had the battle had a different outcome, the Romans no doubt would have taken control of northern Germany and from there moved farther north. The Latin influence that shaped the languages of France and Spain might have done the same to Germany. We might not have the Swedish, Norwegian, Danish, Dutch, German, or English languages today.

As the English historian Sir Edward Creasy wrote in the dramatic style of 1851, "Had Arminius been supine or unsuccessful, our Germanic ancestors would have been enslaved or exterminated in their original seats along the Eider and the Elbe. This island would never have borne the name of England, and 'we, this great English nation, whose race and language are now overrunning the earth, from one end of it to the other,' would have been utterly cut off from existence. . . . It was our own primeval fatherland that the brave German rescued when he slaughtered the Roman legions."

The German/English people, spared from Roman domination, would eventually dominate the landscape themselves, thanks, in no small part, to a plague whose spread was aided by patterns of the weather.

Why the Sun Never Sets on the British Isles

"We saw desolate and groaning villages and corpses spread out on the earth," wrote John of Ephesus. "Corpses which split open and rotted on the streets with nobody to bury them . . . their bellies were swollen and their mouths wide open, throwing up pus like torrents, their eyes inflamed and their hands stretched out upward, and the corpses, rotting and lying in corners and streets and in the porches of courtyards and in the churches."

Around the year AD 541 it began. The first recorded visitation of an illness that would spread throughout the known world, laying waste to cities, leaving so many corpses piled up that there were not enough people left to bury them. Fields were abandoned, leaving cattle to roam freely and crops to go to seed. Those who were spared by the first outbreak of the bubonic plague could only watch, horror-struck, as their loved ones succumbed. The disease came in waves. If it spared you the first time through, it might not be so kind when it came through again a year later. The effects of the disease were gruesome. The sick peered at the world through bloodshot eyes. Covered with pustules, their faces would swell and breathing would become difficult. Some mercifully lost their senses before finally dying of the swelling and high fever. The bubonic plague of the Justinian era ravaged the Roman Empire, killing as many as half of its inhabitants, and changed the balance of power throughout the world.

The original Britons were not the English but the Celts. In the year AD 449, these inhabitants of the British Isles were set upon by an invasion of a barbarous tribe known as the Anglii, or the Angles. They came from the area now known as Schleswig-Holstein and spoke a language that had evolved from the speech of the Frisians, a tribe that lived among the marshy islands off coastal Holland. Their language was the forerunner of English, German, and Dutch.

The Angles, together with the Saxons and Jutes, unleashed terror on the British Isles. "Never was there such slaughter on this island," wrote one chronicler. The native Britons were forced to flee west "as if from a fire." Over the next decades, the invaders spread out and made themselves at home. Even though the Angles were the newcomers, they dubbed the original inhabitants of Britain *wealas*, their word for "foreigner"—from which we get the word "Welsh." This may help to illustrate why the Celts were not particularly fond of their neighbors. The Angles and Saxons in the East and the Celts in the West wanted nothing to do with one another.

"We might expect that two languages . . . living alongside each other for several centuries would borrow freely from each other," wrote the authors of *The Story of English*. "In fact, Old English . . . contains barely a dozen Celtic words. . . . It is as though the English could not be bothered to learn the language of the island they had conquered."

Rather than trade with these despicable characters, the Celts took to the seas and did their business with the people of France, Spain, and the Mediterranean. This would have far-reaching consequences for the Celts and the English. When the plague spread across Europe, it first traveled to Britain in the cargo holds of ships. The result is that the Celtic population in the West, which traded heavily with the people of the Mediterranean, was devastated by the epidemic. The Angles and Saxons were not. This tipped the balance of power on the island and allowed the warriors of the group that spoke the forerunner of English to move into the weakened Celtic regions and colonize them as well. It was the beginning of what would become the British Empire. English colonial influence would spread into Ireland, Wales, Scotland, and then to North America, the Caribbean, India, Australia, and the United States. "The dominoes that fell in Britain over the centuries following the climatic and epidemiological events of the sixth century ultimately changed the world," wrote archaeologist David Keys, "perhaps even more spectacularly than the legacies of most other nations from that period."

What triggered this outbreak that so changed the balance of power in the world? It was, of course, the weather. Some time around AD 530, a dramatic event occurred that blocked out much of the sun's heat for more than a year. Scientists have speculated that it was a massive volcanic eruption or a collision by a comet.

Whatever it was, this event spread dust into the skies and wreaked havoc on the world climate. The effect on East Africa was a severe drought, which ended suddenly with a deluge of rain. The drought killed crops, which unleashed a chain reaction in the ecosystem. Gerbils and mice that fed on the grain died, and the larger predators that normally would eat the rodents also died. As soon as the drought ended, however, increased rainfall brought plant life back at a speedy pace and the fast-breeding gerbils were able to replace their numbers. Because their larger predators took longer to spring back, the rodents were able to multiply like, well, rodents. Under optimal conditions, a pair of gerbils can produce more than a thousand descendants in a year. For a brief period, East Africa was overrun by mice and gerbils.

The gerbils could carry the plague bacteria but were immune to it themselves. Fleas, which fed on the rodents, on the other hand, are not immune to the infection. The African fleas fed on infected rats, became ill, and their guts became engorged with clotted blood. They began to starve. The famished fleas bit any creature they could find, yet no matter how much they bit, they were still hungry. So they bit some more. With each bite they spread the bacteria to a new host. This is how the plague was passed to black rats, also known as ship rats because of their habit of acting as stowaways in cargo holds. The rats traveled along to the port of Pelusium, where the Romans unloaded vast quantities of ivory.

At its peak, Rome imported about fifty tons of ivory a year from Africa. But the plague that spread through the ports virtually halted the ivory trade. Constantinople was hard hit by the epidemic and shrank from a city of half a million to one of fewer than a hundred thousand. From there the illness spread north to France and Britain, and the Angles began their conquest.

The truly dominant force to emerge out of the plague is the English language. Of the twenty-eight hundred or so languages that exist on earth today, only ten are the native tongues of more than a hundred million people. English is the first language of about 350 million: about one-tenth of the world's population. Only Chinese has more native speakers. English, however, has become the second language of choice for the world. Nonnative English speakers now outnumber native speakers two to one. There are more English language students in China now than there are residents of the United States. One out of every seven people in the world today understands or speaks English to some degree. Most of the world's radio broadcasts, books, newspapers, and international phone calls are conducted in the language that evolved from the speech of Germanic tribes who had the good fortune to remain well while their Celtic neighbors were cut down by the bubonic plague. Native speakers of the Welsh language now number fewer than 327,000.

The First Kamikaze

"In Xanadu did Kubla Khan / A stately pleasure-dome decree," wrote English poet Samuel Taylor Coleridge, referring to the Mongol emperor and grandson of Ghengis Khan. Xanadu was actually Shang Tu, the upper capital of that empire, where legend tells us a pleasure dome was built. Here, it is said, Kublai Khan drugged soldiers from conquered nations, gave them a taste of paradise, and then told them that they could return to this heaven on earth only if they served their new emperor well in battle. Most historians believe the pleasure dome was nothing but a myth, created by the conquered people who needed some kind of explanation for their fate.

By the time Kublai Khan took control of the Mongol empire, it stretched to the Black Sea and the Mediterranean. Through years of bloody battles, the Mongols had crushed the resistance in northern China and controlled Bengal and Tibet. Kublai added Korea to his empire. The conquest of Japan was an all-but-inevitable next step in his "heavenly mission."

In 1273 Kublai sent a message to Japanese emperor Kameyama,

which said, in essence, that the emperor should be grateful to be ruled by the Khan, and that he could agree to it now or he would have to agree to it after a ruinous war. Much to Kublai's surprise, the emperor declined the generous offer.

So Kublai sent out an order for a thousand ships to the king of Koryu, a vassal state of the Mongols. The king did not believe he could fulfill such a large order, but Kublai managed to convince him, not with an invitation to a stately pleasure dome but with an army of five thousand armed horsemen. The workers of Koryu did all they could, but they were hampered in their efforts by severe weather and a food shortage. As anxious as Kublai was to take over his new island, he would just have to wait until spring.

In 1274 the Mongol armada, filled with twenty-five thousand mounted soldiers and their horses plus fifteen thousand Koreans, set sail for Japan. They landed and occupied the islands of Tsushima and Iki in preparation for an attack on mainland Japan; on the way, they massacred the local fishing population. On November 20 they proceeded to Hakata Bay, where the Mongols first encountered samurai warriors. The samurai were accustomed to ritualistic battles. They stood in the open and announced their name, ancestry, and victories before challenging a single combatant to what was essentially a duel. The Mongols found this practice quaint and useful. While a samurai was boasting of his parentage, a Mongol horde attacked him. It did not take long for the Japanese to realize that the Mongols were not playing by the rules, and they withdrew to a defensive position, to rethink their strategy. Meanwhile, the Mongols were out in the open, ripe for attack. As night fell, they decided to go back into their ships, which they thought would provide them safety and protection, leaving their Korean vassals behind to be slaughtered by the Japanese. To cover their retreat, they set fire to the Hagosaki shrine.

What they did not know was that a typhoon was headed directly toward them. Only hours after they boarded their ships the storm snapped their masts and flung the vessels left and right, dashing them against the rocks. Horses jumped off decks, and their riders were washed ashore after them. The Mongol soldiers who did make it to shore had the misfortune of being rescued by fishermen from villages that they had burned and pillaged. They did not survive long.

Kublai had never lost a campaign before, and to lose most of his soldiers to the weather, not battle, was almost too much for him to bear. He vowed to sail again with a larger fleet and finish what he had started. He formed the "office of chastisement of Japan" to fulfill this mission.

In the meantime, Kublai conquered the Sung dynasty in the south of China and sent more missives to Japan urging them to surrender without another war. The messengers did not get quite the welcome they might have wished; they were beheaded. The threats did have one effect: the Japanese expected the invasion. They turned their attention to building a shoreline wall at Hakata Bay.

Meanwhile, Kublai Khan was assembling one of the largest invasion forces ever seen. He did this by commandeering every vessel he could find from southern China to Korea. He also put in an order to the Koryu king to build another one thousand ships, and the 1279 conquest of the South China Empire added another thirty-five hundred vessels to the fleet. In 1281 this impressive showing of marine power headed out under thousands of dragon banners. That June they reached Tsushima Island and Iki Island and conquered both before heading to the mainland of Japan. This time, the Japanese were prepared. When the invaders reached Hakata Bay, they were met by an army fighting as a team from behind strong defenses. It was not what they had expected, so the

Mongols and their vassal soldiers withdrew to their ships to prepare for the next assault.

But the Pacific was building its own forcc. Winds in excess of 75 mph (125 kph) were swirling around the dead eye of a typhoon headed directly toward southern Japan. It hit just as the Mongol army was loading onto ships. The impressive fleet was no match for the fury of the waves created by this violent storm. Ships were ripped from their anchors; masts fell. With echoes of the Battle of Salamis, the huge number of vessels would work against Kublai's force. The hulls of wrecked ships broke other ships to pieces and littered the way so the remaining vessels could not escape. It was said that a man could walk across the bay on the wreckage.

Some stunned and waterlogged soldiers made it to the island of Takashima, but most died there. Those who did not drown were easily (and enthusiastically) killed by the Japanese. When the last Mongol general, Chang Pak, died, so did the plan to invade Japan. Kublai Khan died in 1294 and his successor, Timor Khan, had no desire to tempt fate again.

The Shinto priests, and many of the people of Japan, believed these two fortuitous storms were brought about by prayer. In honor of the gods who had protected them, they called the tempest *kamikaze,* an expression that is generally translated into English as "the divine wind." But as the historian Peter Metevelis noted in the journal *Asian Folklore Studies,* the translation fails to capture the full spirit of the Japanese expression. "This translation implies wrongly that the wind is an instrument of the deity rather than the deity herself. . . ." Metevelis suggests that a better translation might be "Deity Wind."

The kamikaze had far-reaching effects on Japanese history. Their good fortune gave them a strong sense of national pride. Japan protected itself behind its fortified seawalls and remained relatively isolated from the rest of Asia.

Paradoxically, many historians believe the Japanese victory led to the collapse of the government. Traditionally, samurai received rewards for their achievements in battle, and these were paid from the spoils of war. Because this war had been on their own soil, there were no spoils. The Buddhist and Shinto priests who believed they had been responsible for the victory were also dissatisfied that they were not rewarded. This led to political turmoil and unrest, which continued for two centuries.

The ill-fated Japanese invasion is also seen by many historians as the beginning of the end of the Mongol dynasty, though it would take almost a century for it to collapse. After nine successors to Kublai Khan, the Chinese revolted in 1368 and the Ming dynasty was formed.

Of course, this was not the last time the term kamikaze was used in war. During World War II Japan would invoke the memory of the storm to create a new kind of warrior, a suicide pilot who turned his aircraft into a weapon against Allied warships.

Loss of the True Cross

If you plan to do battle in a desert, it is best to take water along. In case you need a cautionary tale to illustrate this principle, you may look no further than the battle of Hattin during the Crusades.

The Crusades were a series of battles waged from the eleventh to the thirteenth centuries on behalf of the pope in order to defeat the "enemies of Christ" and to protect the holy sites of Christianity, especially Jerusalem. During two centuries of warring, the various battlefields and disputed territories spread across much of Europe and the Middle East. As a reward for participating in the holy wars, Christian soldiers were forgiven their sins and gained salvation for eternity. Their salvation thus guaranteed, there was little to curb their enthusiasm for plunder and mayhem after a victory in battle. Muslims were not the only targets of the Crusades. Crusaders fought against the evils of pagan Wends; Orthodox Russians and Greeks; shamanist Mongols, Baits, and Lithuanians; and heretics of every stripe including Catholics who questioned the papacy. Occasionally, in their zeal, crusading armies captured

Christian cities that happened to be on their travel route and were in possession of something the crusaders wanted, like gold.

It is safe to say, however, the crusaders did not make friends of the Muslims. In 1099, when they captured Jerusalem, crusaders massacred the entire Muslim population. The result was a holy jihad that lasted for the next two centuries and which, many argue, never really ended. Still, Jerusalem became a Frankish kingdom, and for a hundred years it remained largely unmolested. The greatest threat to emerge to the kingdom was the sultan Salāh ad-Din, known in the West as Saladin, in the late twelth century.

After the death of the emir Nur ad-Din, Saladin had proclaimed himself sultan of Egypt and spread his conquests to northern Africa, Yemen, and Damascus. His goal was to unite all Muslim people. He gathered a huge force of Muslim warriors from various groups; Christians referred to them collectively as Saracens. The Christians and the Saracens eyed one another warily, but a fragile peace existed between them. Saladin and Raymond III, the count of Tripoli, were on good terms. Both were devoted to codes of honor, and they respected each other.

In 1185 King Baldwin IV had died, leaving a young nephew, Baldwin V, as king; Raymond was named his regent. He established a truce with Saladin, and everything seemed to be in balance until the eight-year-old monarch died. A power struggle followed, and in the end Guy of Lusignan became King Guy I. He surrounded himself with a group of barons who were motivated more by greed than by spiritual salvation. The new king's leading adviser was Reynald de Châtillon, who made his fortune by robbing caravans en route to Mecca.

One night, Reynald heard about a particularly wealthy camel train that was soon to pass by the kingdom. Reynald's followers raided the caravan and came back with gold and spices and a beautiful maiden, who, as fate would have it, was Saladin's sister. Reynald was overjoyed. Saladin's sister seemed ideal to command a great ransom, but the sultan refused to pay.

Meanwhile, Saladin's son, Malik al-Afdal, had asked for and been granted safe passage through Raymond's provinces. He agreed to pass through the territory between sunrise and sunset and not to bother the towns. As the Turks were camping peacefully, a group of knights, led by Gérard de Ridefort, attacked. It was a foolish move; the knights were vastly outnumbered by the Muslim warriors, who surrounded and decapitated most of them. Ridefort and three of his knights escaped to tell the tale. The Saracens crossed out of the area before sunset, as they had agreed.

After these two assaults on his family, Saladin vowed to rid the world of Reynald. King Guy was also furious when he heard of the massacre of Ridefort's men, and he assembled a great army of knights to avenge the killing. A treasury that King Henry II of England had donated for the defense of the Holy Land was thrown open and used to hire mercenaries. They assembled a force of twelve hundred knights and eighteen thousand infantry. Guy ordered Heraclius, the Patriarch of Jerusalem, to fetch the True Cross to lead the Christian army into battle.

After a skirmish at Cresson, Saladin's army attacked Tiberias and set it on fire. Next they would take Galilee. Saladin's friend Raymond, who had remained on the sidelines up to that point, could not allow Galilee to be taken. He rejoined his fellow countrymen in the crusade.

The night before July 3, 1187, the Frankish knights were poised to march into the desert that separates Saffuriya from Tiberias on the way to Galilee. Raymond was certain that marching into such an unforgiving landscape in July in heavy chain mail could only spell certain death. Yet because he had not supported Guy in his claim for the throne of Jerusalem, the king was not interested in Raymond's advice. He ordered the knights to mount up and ride into the blazing desert. He reckoned that the journey across the desert would take less than a day, and he didn't want to be slowed down by oxen pulling carts. He ordered that the water be left behind.

Raymond accepted his king's orders and led the vanguard. In the center rode King Guy himself, and behind him Bishops Ruffin of Acre and Bernard of Lydda bearing the True Cross—a sacred relic believed to be the actual cross upon which Jesus was crucified—leading the ranks of twelve hundred knights and seven thousand foot soldiers. The problem was that he had based his calculation of the entire army's speed on how fast a knight could travel at a brisk canter. Not only was such a pace difficult with a thirsty and exhausted horse, but it was also much faster than a foot soldier could possibly travel.

As the limestone reflected the sun from the brutal heat, the knights, under their armor, were red-faced and dizzy. They quickly exhausted the supplies in their water bottles and they would not see another watering hole until they hit the Sea of Galilee. King Guy could have taken a small detour toward the spring of Turan and spared his men, but he was intent on moving forward.

The dehydrated crusaders were sluggish; their pace slowed even more. Their horses collapsed underneath them. Some soldiers simply collapsed and died. Just as they began to spot their salvation—the water of the Sea of Galilee—the Saracens blocked their advance by setting fire to dry brush. The fire created thick smoke that further aggravated dry throats and eyes. From over the blaze, Saladin's men fired a volley of arrows. The crusaders hardly cared. All they could think about was water. Some of the mercenaries used the opportunity to surrender and converted to Islam in exchange for a glass of water.

Raymond knew of a well and suggested changing direction and moving into the hills toward the Horns of Hattin. Saladin's solders, under his nephew, Tawui Ad-Dīn, blocked this route as well, and a fierce battle ensued. The crusaders knew the only thing standing between them and water was the enemy. They fought with everything they had and managed to break through the line. Instead of making immediately for the watering hole, however, the king ordered the soldiers to camp for the night.

"And so, in sorrow and anguish, they camped on a dry site where, during the night, there flowed more blood than water," a nineteenth-century scholar recorded. "That night God indeed gave them the bread of tears to eat and the wine of compunction to drink."

All night the Saracens taunted the Christians, who were nearly mad with thirst. They carried water up to the edge of the camp and, within view of the crusaders, poured it into the sand.

The next morning the sultan sent a messenger who ordered the king and his men to go back to where they came from and never return. The king refused to surrender. Then he ordered his troops to proceed in an orderly fashion toward the well. There was nothing orderly about their procession. They scrambled, knocking one another over, and before they could get to the water, they came face-to-face with a heavy formation of Saracens.

There was no escape. Only Raymond's unit was allowed to slip into the hills. What little strength the Christians had departed them when the Sultan's nephew killed the bishop of Acre and rode off with the True Cross. The True Cross now lost, there was no more reason to fight, and most of the soldiers lay down their arms and waited to be captured. When the fighting ended, only two hundred knights and one thousand foot soldiers remained of the once proud Christian army. The Battle of Hattin was the beginning of the end of Christian supremacy in the Middle East; after this battle every Christian city and castle within the kingdom, except Tyre, fell to Saladin.

Greenland's Vikings

Victims of Climate Change and Cultural Stubbornness

The largest island in the world is also one of the least hospitable environments for human life. Mostly covered by ice and surrounded by a sea dotted with ice floes, Greenland seems an unlikely location for a medieval colony, but the Vikings survived and thrived among the fjords of Greenland. They built farms and churches, and used their settlements as a jumping-off point to explore the coast of Canada—the first Europeans to explore the New World. Between 984 and the 1400s, the Greenland colonies thrived, boasting five thousand settlers at their peak. Then they simply disappeared. The disappearance of the Greenland colonies has intrigued historians and archaeologists for centuries. Modern researchers believe that their disappearance was due to a combination of worsening climate and stubborn Norse refusal to adopt the habits of their Inuit neighbors.

The Norse Vikings built the finest ships of the age. They were streamlined, powered by both sails and oars, and could operate on both shallow waterways and open seas. They dominated the seas, finding trading partners and monasteries to pillage from Britain to

Baghdad. They may even have ventured as far south as Africa. They created the duchy of Normandy in what would later become France; Lincoln and York in England; and the dynasty that ruled Kiev, in Ukraine. Then in the ninth century an especially mild climate known as the Medieval Warm Period made it possible for the Vikings to head west. In about 870 they reached Iceland, where twelve thousand Vikings eventually settled.

"Viking," incidentally, refers specifically to Norse sailors. The word most likely derives from the Old Norse *vik,* which means "bay." A Viking was a type of Norseman, not a nationality. When a Norse sailor got off his ship and settled down on a farm, he stopped being a Viking.

The story of Greenland begins with a passionate Icelandic settler, Erik the Red. When his temper flared, he went—to use an Old Norse word—berserk. Erik had been banished to Iceland for murder. He settled down on a farm, married a Christian woman, and had four children. This pastoral existence did not last long, however, before Erik's temper reared its head again. He fought with a neighbor over a cow, and next thing you know, the neighbor's sons were dead. Iceland's general assembly, the Althing (which still exists today) voted Erik off the island.

Erik set sail toward a then-unknown land that was visible to the west. He found the most hospitable tract of land, a deep fjord on the southwestern coast. Warmer Atlantic currents met the island there, creating a climate similar to Iceland's. Sensing that a name like "Even-Icier-Land" would not induce much tourism, Erik decided to call this new country Greenland. He returned to Iceland in 985 and convinced a group of followers that Greenland was the perfect place to make a new life.

Some twenty-five ships set out for Greenland. Only fourteen survived the trek, bringing 450 new colonists ashore. Technically, as Greenland is geologically part of the Canadian Shield, Erik's family and friends were the first European settlers in North America. News of pastureland available for the asking inspired more settlers. Over the next decade, nearly all the suitable farmland was snapped up. The colonists grew hay to feed imported livestock, such as cows and sheep. They did not eat the animals but relied on them for milk and wool. For food they feasted on seal meat. (It's high in iron, low in fat, and Canadian restaurant critic Barry Lazar recommends a nice red wine with it.)

Although the two Greenland colonies were south of most of Iceland, the climate is more severe because Iceland and Norway are heated by the Gulf Stream, which flows from the south. The current that affects Greenland brings arctic air, with icy winds and fog. They blow drift ice from the north, which frequently blocks the fjords with icebergs—even in the summer.

Calling this island "green" was apparently a symptom of a larger problem the Norse had with naming things. One of the settlements was almost directly north of the other. Yet instead of naming them the Northern and Southern settlements, they called the larger one, with a peak population of about four thousand, the Eastern Settlement, and the smaller one, with a one-thousand-member population, the Western Settlement. This would cause quite a bit of confusion when later Europeans went looking for the settlers. They expected the Eastern Settlement to be on the east coast. It was, in fact, on the west coast.

The early settlers encountered no Inuit. The same warm climate that had allowed the Vikings to expand westward had allowed the Inuit to journey. The warm temperatures melted some channels between the northern Canadian islands that had previously been frozen shut. This meant that bowhead whales, the preferred diet of the Inuit of the region (sorry, no wine recommendations), could migrate into Canadian waterways. The Inuit followed them into Canada, but they returned when the climate shifted back around 1200.

The Greenlanders remained staunchly European in their outlook. They were hungry for information from back home; when ships arrived, they copied the European clothing fashions. (One thing the Vikings did not wear are those helmets with the two horns coming out of them—they were created for opera costumes in the nineteenth century.) They enjoyed active trade with Europe. Their most prized exports were walrus ivory and hides. When the Crusades cut off access to elephants, walrus ivory was the only ivory available. Another luxury item from Greenland was polar bear hide. Polar bear fur coats were rare enough to be a stunning status symbol. One of the most unusual trade items was the baculum of male walruses. This bone, which forms the core of the walrus penis, was just the right size and shape to make an axe handle.

Early in the twelfth century, the Greenlanders decided they wanted a bishop, both to give their church some authority and to cement their ties with European culture. They sent Einar Sokkason, one of their leaders, to Norway to convince the king to send them one. Initially, Bishop Arnald refused the appointment, saying, "I'm no good at handling difficult people." But the Greenlanders sent the king a gift of a live polar bear, which apparently did the trick. Soon they were trading their baculums for church bells and stained-glass windows. By the fourteenth century, the church owned two-thirds of the island's best pastures. The church collected tithes, much of which went to support the Crusades.

When the Thule Inuit people began to immigrate back to Greenland from northern Canada in 1100, they got off to a rather bad start with the Europeans. The Norse called these people *skraelings,* or wretches. One of the earliest references to the *skraelings* describes an eleventh-century incident. "Farther to the north beyond the Norse settlements, hunters have come across small people. . . . When they are stabbed with a nonfatal wound, their wounds turn white and they don't bleed, but when they are mortally wounded, they bleed incessantly." This encounter led the historian Jared Diamond to comment wryly that it "bodes poorly for peaceful relations if you take the first Inuit . . . person whom you see, and you try stabbing him as an experiment to figure out how much he bleeds."

Once they were aware they had company, the Norse became even more protective of their identity as Europeans. There was very little trade between the Norse and Thule, and the Norse refused to adopt some Thule customs that might have spared their lives. The Thule had lived in Arctic regions for centuries, and they knew how to fish and hunt through the ice. They used harpoons and built boats out of sealskins. As the climate cooled, these tools might have been very useful to the Norse. Meanwhile, the scarcity of iron in Greenland meant that the Norse did not have the types of weapons that usually gave Europeans a military advantage over native peoples.

Even in the mild "warm period" in Greenland, survival was far from guaranteed. The growing season was short. There was still frost in August, and the fjords began to freeze in October. There was no natural wood—only the occasional piece of Siberian driftwood that washed ashore. The rest of the colonists' lumber was imported from Norway and later from Canada. Iron was imported to create weapons, tools, and church bells. The Norse in Greenland were completely reliant upon imports of these much-needed materials. The worst thing that could befall the colonists would be a string of especially cold summers that would reduce the output of the farms and prevent sailors from coming ashore to trade. This is exactly what happened.

As the period known as the Little Ice Age swept across Europe, temperatures dropped in Greenland. The longest cold spell stretched from 1343 to 1362. Many of the arctic animals, the mainstay of Norse hunting, migrated to warmer climates, and the Norse were now competing with the Inuit for this food supply. The fields could not produce enough hay to feed the imported cattle, sheep, and goats. Meanwhile, shipments from Europe slowed as drift ice appeared along vital trade routes.

"Now the ice is come from the north, so close to the reefs that none can sail by the old route without risking his life," wrote Ivar Bardarson, a Norwegian priest who lived in Greenland from 1341 to 1364. An increasing number of sailors were not willing to make the death-defying journey for a few walrus bones.

What is more, the demand for ivory was not what it used to be. The Crusades gave Christian Europe new access to elephant ivory, and by the 1400s ivory, by virtue of its availability, was useless as a status symbol and fell out of fashion.

Increasingly cut off from Europe, with poor crops and a dwindling number of animals to hunt, the population of Greenland became very hungry. The Western Settlement was the first to succumb. It disappeared around 1350. In 1361 Ivar Bardarson traveled to the Western Settlement and found it completely abandoned.

In 1397 the last bishop of Greenland died. The pope appointed a new bishop, but from then on he would preside over the colony from the warmth of Rome. Without a resident priest, the authority of the church eroded, and Greenlanders poured into the huge Gardar Cathedral and its church-owned lands.

As the North Atlantic became stormier and colder, traffic to Greenland slowed and then stopped entirely. The last ship to leave Greenland did so in 1410. Now completely cut off from European supplies, the Norse were forced to butcher their cows and even their dogs. Twenty-five years after the last ship, the Eastern Settlement was gone. Yet Europe was blissfully unaware of the Greenlanders' fate. As late as 1600 the pope was still appointing bishops to Greenland.

In 1607 a Danish-Norwegian expedition set out to find the settlement. There was no evidence of it. Of course, they were looking for the Eastern Settlement on the east coast.

The Treaty That Fell from the Sky

Ice hurtling from the sky in the form of hail is one of nature's most dramatic and potentially damaging shows. Hail is made up of spherical balls of ice formed from small frozen raindrops, called graupel, which are recycled through the up and down drafts of a thunderstorm. They keep accumulating new layers of ice until they become so heavy that they fall to the ground. The heaviest documented single hailstone fell on Coffeyville, Kansas, in 1970. It was 5.67 in in diameter and weighed 1.67 lb. In 1925 an undocumented giant stone weighing 2.04 kg (4 lb) crashed into Germany. It was made up of several large hailstones frozen together.

In a more superstitious time, hail was often regarded as an omen from God. Europeans tried (without much success) to ward off hailstorms by ringing church bells and firing cannons. So when a torrent of hail, with stones the size of pigeon eggs, pummeled Edward III's army on the march between Paris and Chartres on April 13, 1360, he took it seriously.

England and France of the 1300s were not nations as they are today. There were fiefdoms of kings, princes, and feudal overlords who inherited property, which they tried to defend and expand with almost constant warring. Edward III's claim to the French throne goes all the way back to 1066, when the British Isles and France were divided between a number of smaller kingdoms that correspond to the current English and French regions. One group of warlike French—the Normans, led by William the Conqueror—landed at Hastings and soon overran English territory. The Normans were actually descended from Vikings—the name Normandy is derived from the word *Norsemen*.

Within two decades of William's conquest, the Norman takeover of English fiefdoms was almost total. Of the 170 tenants-in-chief (barons who held land directly from the king), only two were English. French was the language of the powerful, Latin was the language of priests and scholars, and only commoners used the Anglo-Saxon tongue. One relic of that time is the large vocabulary of the English language. In English, a single concept will often have three synonyms, one derived from Anglo-Saxon, one from French, and a third from Latin (example: king, sovereign, regent). The conquerors and conquered intermarried and after a century it was very difficult to sort out the Normans from the English. Powerful Anglo-Norman nobles often claimed land on both sides of the English Channel and divided their time between the two.

In 1152 Eleanor of Aquitaine married Henry Plantagenet, Earl of Anjou, Duke of Normandy. As a wedding gift, Henry, who would later become England's King Henry II, received the province of Aquitaine, in southwest France. This particular Anglo-Norman union would prove to be quite troublesome to future generations.

In 1204 the Anglo-Normans, under King John, lost control of

their French territory across the Channel, although the military and political victory was not that neat. Because of the interbreeding, it was now very difficult to divide the territory between the "English" and the "French." In 1244 the king of France ordered subjects to choose their allegiance. "As it is impossible that any man living in my kingdom, and having possessions in England, can competently serve two masters, he must either inseparably attach himself to me or to the king of England." They may have chosen allegiance to one side or the other, but that didn't mean they were willing to give up valuable territory.

Which brings us to the start of the Hundred Years War. See if you can follow this: When the French king Charles IV died in 1328, Edward III of England, who was also Duke of Guyenne (in Aquitaine, France) and count of Ponthieu (on the English Channel), considered himself to be the rightful heir to the French throne. Under the law of succession in France, women could not rule, but Edward argued that this did not exclude their male descendents. Edward's mother was Charles IV's sister. Charles had no sons. Therefore, he reasoned, he was next in line. The French assembly disagreed and chose Philip VI, son of Charles of Valois and grandson of King Philip III, as the successor. Edward appeared to accept the decision at first, but eventually returned to reframe his argument with the aid of a large army.

In 1340 Edward gave himself the title King of France. He took advantage of a political conflict in Brittany and Normandy and intervened in a number of important French provinces, which became bases for a military campaign in northern France. The expedition was largely an excuse to loot and plunder. For the next fifteen years Edward's knights roamed the countryside and made short work of any attempts at resistance by the French. In 1346, the English won the battle of Crecy. A year later they captured the strategically important town of Calais. In 1356 Edward's son, known as "The Black Prince," captured Philip VI's successor, John II. By November 1359 Edward's army was marching toward Reims, where French monarchs traditionally were crowned. It seemed the Norman Conquest might be reversed.

Reims proved to be more of a challenge than Edward had anticipated. The siege of the newly fortified city in winter, with troops exhausted by the cold and lack of crops to steal, failed. An attempt to capture Paris that spring also failed. On April 13, 1360, the English headed toward Chartres. The sky became dark, and a storm of great fury began. Lightning hit soldiers and turned their metal armor into giant conductors. Hail said to be the size of pigeon eggs rained on the men and their horses. It struck with such force that when the chain mail was removed from the Duke of Lancaster, iron rings had been pressed into his skin, leaving a bloody pattern. The horses and the common soldiers clad in nothing but leather were killed outright. It was a battered and frightened army that finally arrived at the village of Bretigny.

Edward III and his men had experienced the wrath of God before, which is how they interpreted calamities brought about by the natural world. The bubonic plague of 1348–1351 was an all-too-recent memory. They had learned that when nature got mad at you, it was time to pack it in. Soon thereafter Edward gave up his claim to France and signed the treaty of Bretigny. The French king John was returned for an enormous ransom, and Edward went back to England.

That should have been the end, but it turned out to be just the beginning. The treaty was hardly favorable to the French: It ceded sovereignty of Calais and Aquitaine to Edward, giving fully half of the South Country to England. The French were only biding their time until they were ready to challenge this landgrab and the heavy taxes that came with it. The peace of 1360 would be short-lived.

My Pope's Better Than Your Pope
Lightning and the Great Schism

A bolt of lightning is one of nature's most dramatic shows. Violent, fast, and seemingly targeted, bolts of electricity have commanded a special place in mythology as messages from gods. In ancient Egypt the god Seth created lightning with an iron spear. Zeus, the supreme god of the Greeks, used lightning to influence the outcome of the Trojan wars. The goddess Athena used some of Zeus's lightning to punish the boastful Ajax. In Scandinavia lightning was created by sparks from the god Thor's hammer. The ancient Persians, according to the Greek historian Herodotus, believe "the thunderbolt chastises the insolence of the more enormous animals whilst it passes over without injury the weak and insignificant; before these weapons of the gods you must have seen how the proudest palaces and the loftiest trees fall and perish."

Medieval churches carried church bells with the inscription "Fulgura frango" ("I break up the lightning"). It was believed that ringing the bells could ward off storms. Ironically, bronze bells were actually a great target for lightning strikes, and more than a hundred bell ringers died trying to ward off the very lightning that killed them.

Thus, when a bolt of lightning struck the electoral chamber where the School of Cardinals was debating the election of the pope and the location of the papacy, it was seen as a clear sign of God's will. After more than seventy years, the cardinals officially moved the papacy from Avignon, France, to Rome and chose an Italian pope—then ran as fast as they could. Unfortunately, the French cardinals came to regret their decision and a period known as the Great Western Schism followed in which two, and eventually three, rival popes competed for authority over Christendom.

That the papacy ended up in Avignon was something of an accident. Pope Boniface VIII had not been on the best terms with the French Crown. In 1296 Boniface issued a papal bull forbidding governments to tax the clergy without permission from the pope. This set off a heated battle between Church and State. In 1301 King Philip had a French bishop tried for treason and sent to jail. Boniface responded with another bull, *Unam Sanctam* ("One Holy"), which stated that the pope was the highest authority over human affairs. Temporal power was to be exercised by the king only as the Church permits because God is superior to the temporal world. A year later the pope confirmed Albert of Hapsburg as Holy Roman Emperor and announced that the emperor was overlord of all other rulers, including the king of France.

King Philip was quick to exercise his temporal power. He accused the pope of all manner of special crimes including heresy, blasphemy, and sodomy. He also tried to demonstrate his superiority by sending Guillaume de Nogaret to Italy to stir up a rebellion. The Colonna family in Rome, which wanted a cardinal from its own clan in the papal seat, had already led an uprising against Boniface a few years before. They were more than happy to act again. They kidnapped the pope and threatened to kill him. He was rescued, but the whole episode was too much for him: he died soon afterward and was briefly succeeded by Benedict XI.

The next pope, Clement V, was French. He thought it prudent for the Church to be on better terms with the Crown, so he moved the papacy to Avignon as a temporary measure, in part to convince Philip to drop a posthumous trial against Boniface. The next six popes would reside in Avignon, each planning to move back to Rome. But one way or another, each ended up staying.

The French papacy was not a hit with everyone. The English, in the midst of the Hundred Years War, especially resented tithing to France. The opulence of the papal court at Avignon had also alienated many Catholics. The Italian poet Francesco Petrarch wrote to a friend: "Now I am living in France, in the Babylon of the West. . . . Here reign the successors of the poor fishermen of Galilee; they have strangely forgotten their origin. I am astounded, as I recall their predecessors, to see these men loaded with gold and clad in purple, boasting of the spoils of princes and nations; to see luxurious palaces and heights crowned with fortifications. . . ."

The last of the officially recognized French popes was Gregory XI, in 1371. Italy was torn by rebellions led by Catholics who wished to liberate themselves from foreign papal rulers. Gregory's first reaction was to send in some Christian soldiers, led by Robert of Geneva. This violence led to more violence, and thousands died in a terrible slaughter. Gregory decided that the only solu-tion for a lasting peace was to once and for all move the papacy back to Rome. Although his health was fragile, he set sail on October 2, 1377. En route the pope faced stormy seas. "Angry waves tossed the ships from side to side; the sails were torn, the ropes broke, the anchors dragged, and the terrified sailors feared the ships would be wrecked," wrote historian Guillaume Mollat.

The difficult crossing took its toll on the pontiff's health. He was, in Mollat's words, "unable to withstand the rigours of the

Roman climate." Just as the European congress was meeting to work out the transfer of Church power to Rome in March 1378, Gregory XI died. Things got really messy after that. According to ecclesiastical law, the election of a new pope must be held where the last pope died.

At the conclave were sixteen cardinals: the majority—eleven—were French; four were Italian; and one was Spanish. As they assembled for the conclave in Rome a great storm was brewing—metaphorically, but also in reality. The skies turned black; rain began to fall. Meanwhile, crowds of anti-French Catholics surrounded the hall. They chanted "A Roman pope! We will have a Roman pope!" Some of the protestors had broken into the belfry of St. Peter's and rang the bells continuously. Members of the mob made it fairly clear that if the next pope was not Italian, the French cardinals might not make it out of the hall alive. As if on cue, a dazzling bolt shot from the sky and sizzled through the hall, splitting some of the furniture and setting it on fire. This could not be a good omen.

All things considered, it made fairly good sense for the conclave to select an Italian. He was Urban VI. But Urban was not to be the unifying figure the Church might have hoped for. Before his election he had been quiet and modest, but once in a position of power, he showed some unpleasant colors. He had a personality better suited to a modern-day talk-show pundit than pope. His favorite expressions were "Hold your tongue!" and "You have talked long enough!" He was quick to call cardinals fools and liars. There were whispers that Pope Urban was mad. Nowhere were the cardinals more unimpressed with this pope than in Avignon.

The French cardinals regretted their decision and the circumstances under which they had made it. The omen in the lightning may not have been that they needed to select a Roman pope after all. God might have been warning them that they were about to make a mistake.

The French cardinals declared that their original decision was invalid because it was made under duress from the mob (and the weather). They called another conclave and chose a new pope, Clement VII. The Swiss-born pope moved his court back to Avignon. Urban, surprisingly enough, did not volunteer to retire.

Now the Catholic populace was at odds over which pope to obey. Supporters of each pope labeled the other group's pope a false pope and called him Judas Iscariot. Both sides claimed that acts performed by the other were invalid. Children baptized under a false pope were not truly baptized, those married were not really married, and those who died and received last rites under a false pope were doomed to hell for eternity. When seven of Urban's cardinals asked him to step down to end the rift, he had them executed.

Upon Urban's death, Clement expected he would be welcomed in Rome as the true pope—but he was not. Urban's faithfuls elected a new Roman pope, Boniface IX. Two rival popes continued to reign until 1409, when the Council of Pisa was convened to settle the matter. Somehow at the end of the gathering the world ended up with three popes. (The council had elected a new pope, Alexander V, to replace the fighting popes, but they both refused to step down.) The matter would not be resolved until 1417 when all of Western Christendom finally agreed on Pope Martin V.

The Mud That Made England

If William Shakespeare is to be believed, the Battle of Agincourt was won by the power of rhetoric alone. The young King Henry V, vastly outnumbered by the French enemy, stands before his men and utters a speech that would inspire even the most ardent pacifist to take up arms for the good of his country.

> And Crispin Crispian shall ne'er go by
> From this day to the ending of the world
> But we in it shall be remembered—
> We few, we happy few, we band of brothers . . .

Off they went to die in glorious battle, six thousand English against sixty thousand French. But against all odds, this band of brothers, tired but renewed by their king's leadership and their love of country, took the day and won the battle. In one sense Shakespeare's Saint Crispin's day speech is true. Thanks to this fiery speech, the Battle of Agincourt stands out as one of the great moments of English history. A 1944 film version of *Henry V* starring Laurence Olivier resonated with British audiences who were facing down Adolf Hitler with only Winston Churchill's rhetoric to urge them on.

The words, of course, were Shakespeare's, not Henry's, and they were written two hundred years after the fact. Shakespeare got a few details wrong—for example, the name of the city where the battle took place was Azincourt, not Agincourt. The real Henry had little of the charisma or oratorical genius of his theatrical counterpart. The unlikely English victory had much more to do with the customs of French nobility and some rainy weather.

The campaign, which ended in the battle of Azincourt, came about when Henry claimed he was the rightful heir to the French throne; and unless the French gave in to his demands, he would attack France and take it back by force. Henry's claim was fairly dubious, but it was not quite as crazy as it might seem if the current Queen Elizabeth suddenly announced she was the rightful ruler of France. Back in 1415 the concepts of "English" and "French" as we now understand them were quite different. In fact, our sense of the English and the French as separate peoples was largely a consequence of the Hundred Years War. The reign of Henry V toward the middle of that century-long entanglement and the Battle of Azincourt helped give the people of his nation a sense of being "English."

Henry V was born in 1387, the eldest son of Henry IV. Raised by his uncle, Richard II, Henry was the first English monarch since the Norman invasion who learned to read and write principally in English. Henry's argument supporting his claim to the French throne was flawed at best. He began with Edward III's old, and unsuccessful, claim to be the rightful possessor of the French crown. Since Henry was the king of England, he should rule France. The only problem is that neither Henry V nor his father, Henry IV, were Edward's descendents. No matter—a war with

France was sure to be supported by his subjects regardless of the logic behind it.

In 1415 the twenty-seven-year-old king invaded France with an army of about thirty thousand men and captured Harfleur. The defenders of Harfleur had been weakened by an outbreak of dysentery, which was soon passed to the invading English. The effects of brisk September nights and disease ravaged Henry's army and cut it in half. The weary group would be no match for the French, and Henry knew it. He decided to retreat to his fortress in Calais and regroup. About nine hundred men-at-arms and five thousand archers started the trek. The feverish men marched 260 miles in seventeen days, often in the rain. With supplies cut off, they ran short of food. Toward the end of the march they were eating nothing but dried meat and walnuts.

Had the French simply allowed this youthful monarch to limp home with a sick and exhausted army, we might be talking about the Sixty Years War. The entire mission would have been an embarrassing failure. Instead, the French cut off Henry's escape route and prepared to end his campaign decisively. With flight no longer an option, his only recourse was to stand and fight. French overconfidence, and some well-timed rain, would provide Henry with a shocking victory that would be a great boost to English national pride.

If the French had gone and looked for geography that would negate all of their advantages, they couldn't have done better than the field at Azincourt. Henry's men occupied a narrow field hemmed in on either side by brush. The French could attack only from the front. What is more, the cramped space made large-scale maneuvers almost impossible. In this narrow field, France's numerical superiority actually worked against them. On the eve of Saint Crispin's day a heavy rain fell, and the newly planted farmland of Azincourt was transformed into deep mud. Servants were sent to place logs on the ground to take the weight of the horses as the armored riders got into their saddles.

The rain fell on the English as well as on the French, and the muddy field slowed soldiers on both sides, but there was a crucial difference in the way the two sides fought that made the mire a deadlier problem for the French. The English relied mostly on their archers, who fired from a distance, with only a small band of men-at-arms. The men-at-arms were arrayed in three central blocks linked by projecting wedges of archers. The archers, many of whom were criminals who had escaped jail by becoming soldiers, were not bound by the same code of chivalry as were the knights. The French noblemen considered them to be ill-bred and not a great threat.

The only way that Henry could possibly take the day was if the French were to march onto the field before the ground had a chance to dry: so he had his archers fire to provoke the French. The large contingent of heavily armored knights rode toward the English line on horseback. As they headed toward their enemy, the battlefield narrowed, and they were forced together as if in a funnel. Even on a dry day, the narrowed field would double the density of troops from two people per square meter to four people per square meter, which reduced the advancing speed by 70 percent. The muddy field slowed them even more, creating a bottleneck.

When the cavalry made it to the other end of the field, they discovered that the archers had planted themselves behind long stakes sharpened to points. After a few horses were impaled, the remaining cavalry turned around, creating gridlock. Then the English archers let loose with a barrage of arrows. As many as forty thousand arrows came down on the French cavalry every minute. Most were deflected by the armor, but the horses were not so lucky—the injured beasts reared and threw their riders. Fallen horses and men made the battlefield even more crowded.

The French men-at-arms, weighed down by sixty pounds of armor, were so crowded that they could not move. Some literally could not lift their arms to fight. When one soldier fell, he brought down the men around him. Many French soldiers died by suffocation—crushed by their comrades, they drowned in the mud. As the bodies piled up, the advancing French columns could only try to climb over them. They were, as contemporary writers said, "building a wall of dead knights."

The archers, now out of arrows, climbed out onto the field and attacked the prostrate knights. An anonymous fifteenth-century chronicler recorded: "When their arrows were all used up, seizing axes, stakes, and swards and spear-heads that were lying about, they struck down, hacked, and stabbed the enemy. For the Almighty and Merciful God, Who is ever marvelous in His works and Whose will it was to deal mercifully with us . . . did increase the strength of our men. . . ."

While the archers were thus occupied, looters attacked the thinly guarded wagon train that carried supplies and the king's war chest. As part of Henry's reserves secured the wagon train, a third line of French horsemen cut through the archers while the English knights were busy taking prisoners. Henry was afraid that if his soldiers turned to fight the cavalry, they would be attacked from the rear by their prisoners. So he gave an order that was completely contrary to the rules of chivalry: He ordered his knights to kill their prisoners. They were shocked and refused, partly because of their code of honor but also because it was the custom to ransom captured knights for gold. The archers had no such code, and no such financial motive. A detail of two hundred marched the prisoners to the slaughter. They cut the throats and bludgeoned some of the greatest nobility of France, sparing only those who would command the highest ransoms. The killing continued until the third French cavalry line withdrew. This is one episode that William Shakespeare failed to chronicle. Henry's disregard for the rules of warfare in killing the prisoners would add fuel to hatred of the English, and certainly did little to shorten the war.

When it was all over, the English had two thousand prisoners left. They were taken to London and ransomed. (Some, like the Duc d'Orléans, remained unransomed. He was held prisoner for twenty-five years.) The field of Azincourt was soaked with French blood. More than ten thousand men-at-arms lay dead, and the ranks of the nobility had been reduced by fifteen hundred. Many of their bodies lay naked, their armor stripped by looters.

From the French perspective, the battle of Azincourt was not particularly significant. It was just one of many battles won or lost in a medieval conflict that lasted more than a century. Their unifying figure would be Jeanne d'Arc, who reversed English rule of French kingdoms in the siege of Orléans in 1429. Her death at the stake made her a martyr and helped cultivate a sense of French national identity as we know it today. For the English, however, the English-speaking Henry V at the battle of "Agincourt" unified them as a people.

As Geoffrey Elton, author of *The English*, put it: "All these ups and downs had their effects on the history of a nation which more than once shifted from arrogant exaltation to deeply resentful despondency and back again. The one continuous effect lay in the confirmation of national self-identification: the English, through all their social ranks, knew that they were wonderful, and to the annoyance of visitors from abroad kept saying so."

William Shakespeare's poetic retelling of the events ensured that Saint Crispin's day would be remembered throughout the English-speaking world long after most of the events and characters of the Hundred Years War had been forgotten.

The Fog of War

In a civil war, it is not easy to tell a friend from a foe. This was nowhere better demonstrated than at the Battle of Barnet, a significant engagement in the British Wars of the Roses. Thanks to a thick blanket of fog, confusion ruled the day. A large number of the casualties were inflicted by allies, and with tenuous alliances and family betrayals fresh in everyone's mind, the battle disintegrated into chaos.

The fight for the English crown known as the Wars of the Roses traces its roots, once again, back to Edward III. Two families claimed they were the rightful heirs to the throne through descent from different sons of Edward III.

Henry V (of the house of Lancaster, wearers of the red rose) did not have long to enjoy his victory at Azincourt. He died in 1422, leaving a sole heir, nine-month-old Henry VI. Whether the boy had the personality to be a leader couldn't possibly be apparent, but it didn't matter. He was the king's son—that's all there was to it. During his childhood Henry's uncles duked it out for control over the baby king's government.

Young Henry grew up to be shy and quiet. He hated conflict and war and spent most of his time at his studies or giving away more money to charity than he actually had. All laudable qualities for, say, a monk. For a king in fifteenth-century England, they were not useful at all. The retiring regent chose for his wife a strong-willed Frenchwoman, Margaret of Anjou. She had the commanding personality that he lacked, but she created a number of enemies who did not admire such traits in a woman.

During Henry's reign England lost French territory to Jeanne d'Arc and Charles VII. The country was deeply in debt. What made it all worse was that Henry VI suffered from mental illness. Had Henry been a competent leader, the house of York (they of the white roses) would probably never have sought the throne. After all, the Lancastrians had occupied the throne since 1399. But Henry in his madness was completely unable to perform the duties of a king. So Richard Neville, the Earl of Warwick (he would come to be known as the Kingmaker), installed Richard Plantagenet, the 3rd Duke of York (a descendant of Edward III) to keep an eye on the nation. When Henry recovered, he wanted his throne back. Even more insistent on the matter was his wife, Margaret. Both sides rallied their supporters, and the next thing you know the English were at war with the English.

We will not go through each of the battles and the various turns of fortune for the Lancastrians and the Yorkists, which make the plotlines of the soap opera *One Life to Live* seem downright linear. We'll just flash-forward to 1465. Henry VI has been deposed and imprisoned in the Tower of London. He had been led to his cell by the Earl of Warwick, who was then a close adviser to the man who took Henry's place, the former Duke of York, King Edward IV. As Edward was celebrating his new position, Warwick sailed for France, as it was the tradition for English kings to marry French princesses, and King Louis XI of France had an available sister-in-law. Such unions were supposed to ensure peace between the nations. It hadn't worked that way in the past, but heck, it had to work eventually.

While Warwick was off looking for a suitable bride, however, the king married the widow of a Lancastrian in secret. The new queen, Elizabeth Woodville, became a royal matchmaker, arranging politically useful weddings for her family—often to the detriment of the Earl of Warwick's family. Finally, Edward announced a pact with Burgundy, which included the marriage of his sister, Margaret of York, to Charles the Bold. This undermined Warwick's relationship with the French king. Warwick was furious, and the Kingmaker decided it was time to unmake the king. With the help of his former enemy, Margaret of Anjou, and King Louis, he deposed Edward and put Henry back on the throne. Warwick's younger daughter, Anne Neville, was married off to Henry's teenage son, Edward, the Prince of Wales. If Henry could hold on to the throne, Anne would one day become queen. (Still with me?)

Thus we come to the Battle of Barnet, where the former king, Edward, and his former right-hand man went to war against each other. Both sides included soldiers who were now allies but who had at various times been foes. It was Easter Sunday 1471. Warwick's army, camped near Barnet, numbered about nine thousand, slightly more than Edward's Yorkist force. The right wing of Warwick's forces was commanded by his brother-in-law John de Vere, 13th Earl of Oxford (husband of the Earl of Warwick's sister); his brother, John Neville, 1st Marquess of Montagu (Richard Neville's brother), led the center; and Henry Holland, 3rd Duke of Exeter (no relation) led the left. Warwick would lead from the center.

Edward's battalions arrived by night, and in order to be ready to fight in the morning, he ordered that they make camp close to the Lancastrians. Edward led the center and Lord Hastings brought up the left. Commanding the right wing was Edward's eighteen-year-old brother, Richard of Gloucester. Richard and Edward were the sons of Richard of York and Cecily Neville, the Earl of Warwick's aunt. The earl had, in fact, been in charge of Richard for a time when the latter was a boy.

Somehow the two armies ended up a little closer than they intended—only 200 yards (182 m) away from each other. Warwick, assuming Edward's men were at a reasonable distance, ordered his cannons to bombard Edward's camp throughout the night. As the cannon sailed harmlessly over their heads, Edward ordered his guns to remain silent so as not to give away his position.

The field of battle was surrounded by marshy ground, which created a dense ground fog. By dawn's early light, the soldiers of the two sides could barely see one another. Between four and five o'clock, each line moved forward to where they imagined the enemy must be. It was only when they clashed that it suddenly became apparent that each army's right flank overlapped the other's left. The left flanks on both sides were attacked from behind—something completely unexpected in warfare of that time. Initially, it was Warwick's advantage. Edward's left flank retreated, with the Earl of Oxford's battalion in hot pursuit; and it was said they did not stop running until they reached London. King Edward's right flank should have had a similar advantage over the Lancastrians, but Gloucester's men had to traverse a deep depression, which is now called Dead Man's Bottom.

Fortunately for Edward, the same fog that allowed them to march into disaster hid the carnage from the bulk of the troops. The Yorkists had no idea how badly they were outnumbered, and they fought fiercely against the center of Warwick's army.

This is when the fog of war shifted entirely in favor of Edward. The Earl of Oxford, having thoroughly routed Hastings's men, had reassembled and returned to the battle. Through the haze they could not see that the front line had changed direction. Instead of charging into Edward's army, they were coming straight at one of their own divisions, under the Marquess of Montagu's command. Montagu could see enough to know that a large force of cavalry was racing toward him, but through the gray skies he mistook the emblem of Oxford's men (a star with streams) for that of Edward (a sun with rays). Thinking it was Edward's cavalry attacking him, he ordered his archers to open fire.

When Oxford's division found itself under attack by Montagu's, their first assumption was that Montagu had changed loyalties midbattle. They began to shout "Treason! Treason!" The earl and his men fled the field of battle, reducing the Lancastrian forces. What is more, Warwick's forces had been thrown into utter confusion, with soldiers no longer sure who was on their side and who was fighting against them.

By eight in the morning, Montagu was dead, Oxford had fled, and Exeter was missing. Warwick, who had been fighting hand-to-hand, ran back for his horse. The lines of battle had moved quite a ways from their original location, and Warwick had a great deal of ground to cover to get back. When Edward learned of Warwick's plight, he ordered some of his men to ride out and save the earl, but they were too late. When they got to him, they found a lifeless body stripped of its valuable armor, with a knife thrust through the eye.

The Battle of Barnet was the beginning of the end for the Lancastrians. The next battle, Tewkesbury, would claim the life of Prince Edward, heir to Henry VI. Shortly after the battle, Margaret of Anjou was captured and eventually ransomed by King Louis. Henry VI died within a few hours of King Edward's triumphant march back to London. Officially, he died of "melancholy," but the true cause was most likely murder by Richard of Gloucester.

Anne Neville did become queen in the end. After Henry VI's death, his son's widow was a much sought-after prize. Richard of Gloucester's brother George, who wanted the Neville fortunes for himself, whisked her away, but Richard stole her from George's house and married her. The Earl of Gloucester went on to become King Richard III.

Lost Siberians

As every schoolchild knows, Christopher Columbus discovered America. Of course, he discovered it in much the same way a modern tourist can be said to have discovered a little out-of-the-way bistro: the locals already knew about it.

The New World was not the sparsely populated wilderness that some history texts would have you believe. In fact, the historian William McNeill estimates the indigenous population of the land we now call America was one hundred million in 1492—when Columbus sailed the ocean blue—while the European population numbered only seventy million.

The natives of what is now New England were not nomads; they lived in towns and villages. They were farmers and skilled craftspeople and had technology that many historians believe rivaled that of Europe.

So how were the Europeans able to colonize this "new" continent so rapidly and completely? It can all be traced to the bitterly cold winds of Siberia. To explain, we'll have to go back to the very beginning.

There was a big bang. On second thought, perhaps we don't need to go to the very beginning. Let's begin our story with a solar system that already exists. Earth has been created with a single satellite and a gravitational pull that allows the planet to hold on to a dynamic and life-supporting atmosphere. Organic life has already flourished. One-celled organisms have become more complex, left the seas, and evolved into mammals. The dinosaurs have come and gone, and human beings are starting to take their place as a dominant species. About forty-five thousand years ago, humans began to develop speech and technical skills. They made tools. While much is made of the invention of the spear and the wheel, for our story there is a technological innovation that was even more important: the invention of the needle and thread.

During the late ice age, the temperatures in what we now call Russia were bitterly, bone-chillingly cold. Modern-day Siberia is a tropical paradise compared to the landscape of the late ice age. This Siberia was covered in ice and glacial lakes. The northerly winds blew frigid glacial dust across the plains. This unforgiving environment was uninhabitable by humans until someone came up with the idea of an eyed needle. This little device allowed people to stitch pieces of fur together. They could combine fur from several animals to make form-fitting coats, hats, and boots.

These ancient ancestors were the first to layer their clothing, a habit that keeps those in the northern climes warm to this day. Next time your mother nags you to wear layers in the cold, you can ponder the fact that her mother, her mother before her, and so on, did the same, and that this noble tradition goes back a good thirty thousand years.

Of course, even the best layered clothing can do only so much. As the climate fluctuated across the tundra, humans moved with it. As it got warmer, they moved farther north; if temperatures fell, they would migrate south. This type of nomadic existence allowed the tundra dwellers to survive. It also spread groups of people over enormous territories. As the ice age gave way to warmer temperatures around 13,500 BC, small bands of hunters moved into the extreme northeastern part of Asia. Some of them wandered into a piece of real estate that no longer exists, the land that archaeologists call Central Beringia. This land bridge connected Asia and North America across what is now the Bering Strait.

Even today the gap between Alaska and Siberia is small enough—only 2.5 mi at its closest—that deep freezes sometimes make it possible for people to walk from one side to the other. Of course, what the atmospheric climate makes possible the political climate has generally not permitted, as John Weymouth of San Francisco discovered in 1986. When he walked across the frozen Bering Strait from the United States into the USSR, Weymouth found himself the focus of an international incident. After two weeks of interrogation and negotiations between the U.S. State Department and the Kremlin, the wanderer managed to convince two governments that he was not a defector or a spy, just a curious guy who thought it would be fun to walk to another continent. He was finally sent back to America in a military helicopter.

Fifteen thousand years ago, there was no Russia, no United States, no border, and—this is the important part—no Bering Strait. The first human residents of the American continent were not setting out on a deliberate journey of colonization. They were not so much Native Americans as Lost Siberians. A few at a time, these Asians moved east as they hunted and looked for food. The climate, the plants—everything in the adjoining land was familiar. The steppe dwellers migrated east and south and filled the heretofore unpopulated land, and eventually Central Beringia disappeared beneath the sea. By the time Europeans rediscovered their relatives in the 1400s, they no longer recognized them. Which brings us back to where we began—1492, when the indigenous population of the land we now call America numbered one hundred million.

Although the Europeans had a slight technological advantage in the form of steel weapons and guns, it was a secret weapon that allowed them to dominate the Americas, a weapon the Europeans did not even know they possessed: germs.

Most of mankind's illnesses up to that point were bred in earth's warmer climates. As humans moved into frozen lands like Siberia, microbes that live outside human hosts for part of their life cycle didn't stand a chance. The Native Americans' Siberian past meant they never developed immunities to the disease-causing germs that plagued Europeans. When they entered a virgin continent, the American natives left disease behind.

In the 1600s, before the Pilgrims landed in Massachusetts, British and French fishermen started trolling the New England coast. Occasionally they would come to shore and interact with the natives. These unremarkable encounters would prove deadly—within three years, a plague had wiped out between 90 and 96 percent of the inhabitants of costal New England. By comparison, the Black Plague killed perhaps 30 percent of the population of Europe. Whole towns lay in waste. There were so many dead that there was no way to bury them all. The disease-ravaged mourners found themselves in no position to fight off European invaders. In fact, one of the reasons the Wampanoags were warm to the Pilgrims at Plymouth was that their tribe was so weakened by illness that they were afraid of being attacked by neighboring tribes to the west and they sought allies to protect them. European colonization was swift because the settlers, in many cases, simply moved into abandoned Native American villages and farms. Mexicans speak Spanish today largely due to this same effect. When the Spanish marched into what is now Mexico City, they found an

Aztec population ravaged by smallpox. There were so many bodies that the soldiers had to walk on them. The Spanish were largely immune to the disease.

As the Europeans settled in, they brought even more illness. They stocked their farms with domesticated animals that were not native to the region: sheep, goats, cows, and pigs. The animals carried streptococcus, ringworm, anthrax, and tuberculosis, all of which could be passed on to humans. Historians have recorded that between 1520 and 1918, there were as many as ninety-three epidemics—including bubonic plague, measles, influenza, tuberculosis, diphtheria, typhus, cholera, and smallpox—among native populations.

In those days, neither the Europeans nor the Native Americans knew about germs. They simply saw that one group of people was being destroyed by an epidemic, while another was left standing. Many natives, in the face of such devastating loss, came to believe their gods had betrayed them. There are historic accounts of native shamans destroying the sacred objects of their tribes. Some converted to Christianity; others killed themselves. By the time the natives were able to regroup and challenge the invaders, the Europeans were well entrenched.

The Europeans, meanwhile, took the plagues as a sign that God was on their side, which reinforced the belief that the land was theirs for the taking. John Winthrop, the governor of the Massachusetts Bay Colony, called the native plague "miraculous."

By the 1700s the mechanisms of disease were starting to be known. In 1721 the Reverend Cotton Mather of Boston inoculated 240 people with smallpox fluid to successfully ward off the disease. This knowledge also created a more calculated effort to benefit from the effects of disease among the natives. In 1763 Jeffrey Amherst, governor general of Canada, came up with a plan to "extirpate this execrable race." He "contrived to send the smallpox among the disaffected tribes." He presented two chiefs with "gifts" of blankets from a hospital smallpox ward. The land that was once occupied by the Norwottuck tribe is now known as Amherst, Massachusetts, in Amherst's honor.

"The archetype of the 'virgin continent' . . . has subtly influenced estimates of Native population," wrote historian James W. Loewen. "Never mind that the land was, in reality, not a virgin wilderness but recently widowed."

The hostile climate of the Asian north provided centuries of unprecedented health to the indigenous people of the American continent, and ironically this good health was to be their undoing. Thus the glacial winds of Siberia paved the way for American society as we know it today.

Which Witch Did This?

Quick—prove that you are not controlling the weather. Can you do it?

Just such a question was put to countless people, mostly women, in the fifteenth through seventeenth centuries. The wrong answer was enough to cost them their lives. The only catch: there was no *right* answer.

The sky over Wiesensteig, Germany, on August 3, 1562, was an ominous black, even though it was the middle of the day. Torrents of rain fell on homes and fields, destroying roofs and windows and soaking crops. Then came the hail pounding down with a ferocity that had not been seen for a hundred years. It pelted an area of several hundred square kilometers. The next day, agonized farmers found their horses and cows lying dead. The trees were barren of leaves. Birds lay dead in the fields alongside the ripped-up remains of what should have been the harvest. What could have caused such unnatural devastation? The scientific answer is the climate change of the period meteorologists call the Little Ice Age. The only answer the distraught farmers could come up with was witchcraft.

Within a couple of days several women had been arrested by the Lutheran justice Ulrich von Helfenstein. Six of the women, who refused to admit to their sorcery, were executed. Others said they had repented and begged for mercy. To show their good faith, they cooperated with authorities and claimed to have seen women from the town of Esslingen, thirty miles (40 km) away, at their witch gatherings. Thus, the witch hysteria spread. Pastor Thomas Naogeorgus took up the charge, but unlike in Wiesensteig, which had suffered so much from the hail, in Esslingen those who cried witch were in the minority. The Esslingen city council warned Naogeorgus to scale back the rhetoric. Three women were arrested for sorcery, but they were quickly released and the pastor was fired. He died a year later.

Back in Wiesensteig, the minister was horrified to hear that his fellow witch-hunter's cries were falling on deaf ears. He could do nothing to stop the devil in Esslingen, so he did the only thing he could: he redoubled his efforts at home. Helfenstein arrested and executed another forty-one women that autumn, and in December, another twenty were sentenced to death. By the end of the year, sixty-three witches had been burned at the stake.

Historians and archaeologists disagree as to exactly when the Little Ice Age occurred. Some scholars put the beginning at 1300, others at 1450. Some say it ended in 1770; others say it stretched until 1860. Most place the most severe weather between 1570 and 1630. What they all agree on is that the weather was volatile, with extremely low temperatures: Dutch canals froze over; ships, if they were able to leave port at all, encountered hazardous icebergs; subsistence farmers faced starvation. Then it would be uncharacteristically warm, and then plunge again into a deep freeze. By 1500 European summers were about 7°C (44°F) cooler than they had been during the Medieval Warm Period.

As crops failed and communities found themselves on the

verge of starvation, frightened populations tried to regain some sense of control over their lives. If they could find someone to blame, then they would have someone to prosecute and the problem would be "solved." Thus, witchcraft trials occurred with increasing frequency. During the 1430s, the first systematic witch hunts happened in parts of Switzerland. Initially, the Church refused to accept the idea that witches were fiddling with the weather. But in 1484 Pope Innocent VIII had a change of heart, and he issued the papal bull *Summis desiderantes affectibus*, which said that witches *could* create foul weather. The principal text outlining the proper treatment of witches, the *Malleus malleficarum*, was published in 1484 at the pope's request. The *Malleus maleficarum* was compiled by the Dominicans Johann Sprenger, dean of the University of Cologne, and Henrich Krämer, a teacher of theology at the University of Salzburg and the Inquisitor of Tirol, Germany. Published in twenty-eight editions between 1486 and 1600, its influence was vast. It described the characteristics of witches and the accepted methods of interrogation (read "torture") and punishment (read "execution"). One chapter was called: "How They Raise and Stir Up Hailstorms and Tempests and Cause Lightning to Blast Both Men and Beasts." Witch trials became epidemic in parts of central and southern Europe, with changing the weather as a common charge.

This was the mood in Germany as the weather took a turn for the worse. The winter of 1561–1562 brought heavy snow to Germany. When this melted, fields were washed away, and livestock starved or withered from illness. Prices for food soared, putting it out of reach of the poorest people. Many churchgoers could only assume it was the work of an angry God, furious at man's transgressions. Witch scares were usually not imposed from above. They swelled from the ground up—peasants who were hard hit by an unnatural event demanded that the powers that be do something. The authorities often felt powerless to contradict the will of the crowd.

As a writer from the Franconian town of Zeli wrote in 1626: "Everything was frozen which had not happened as long as one could remember. And it caused a big rise in prices. . . . As a result pleading and begging began among the rabble, questioning why the authorities continued to tolerate the witches and sorcerers destruction of the crops. Thus the prince-bishop punished these crimes."

Throughout the 1560s witch scares flared up in various parts of Europe. There were small, individual accusations across the continent, but large-scale trials were usually the result of large-scale "unnatural" events, that is to say, strange weather. Witches burned after crops failed in 1570. The pious were rooting out witches from Spain and Portugal all the way to Russia and Scandinavia. Between 1580 and 1620 in the Bern, Switzerland, region alone more than one thousand people were executed as witches. Men as well as women were accused and executed by hanging, burning, drowning, or other more creative means.

Some of the largest witch hunts in European history took place in Lorraine and Trèves, France, where between 1581 and 1595 an estimated twenty-seven hundred people were sentenced to death for sorcery. Johann Linden, canon at Saint Simeon in Trèves explained the reasoning behind the prosecutions: "During the whole period [Archbishiop Johann VII von Schonenberg] had to endure with his subjects a continuous lack of grain, the rigours of climate and crop failure. Only two of the nineteen years were fertile. . . . Since everybody thought that the continuous crop-failure was caused by witches from devilish hate, the whole country stood up for their eradication."

In their zeal to protect themselves from the evildoers and make a panicked public feel safe, judges relied on rumors, gossip, and statements extracted through torture. The targets of these inquisitions said whatever their questioners wanted to hear in order to avoid pain. To save their own lives, witches would feign remorse and conversion and become witnesses against the devil. They named conspirators, who would be pulled into court in an ever-widening circle of accused devil-worshippers.

One of the most outspoken critics of the trials was the German priest Fredrich Spee, a prison chaplain who ministered to witches on death row.

"I will state under oath, I have not led any woman to the stake who, with all things considered I could prudently state was guilty," he said. "The only reason we are not all sorcerers is that torture has not yet touched us."

Torture seems to have touched villages whenever cold did. Historian Wolfgang Behringer studied the European witch trials and found a correlation between the peaks of witch prosecutions and "cumulative sequences of coldness" in the years 1560–1574, 1583–1589, 1623–1630, and 1678–1698. From 1730 on, the climate became more stable, and so did the general mood. Isolated witch trials continued in Central Europe until the 1770s, but nothing on so grand a scale as the heyday of the Little Ice Age.

"So it is more than a mere metaphor," wrote Behringer, "that the sun of the Enlightenment ended the era of witch-hunting."

A Protestant Wind Destroys the Spanish Armada

The defeat of the Spanish Armada in 1588 has been called one of the most decisive battles in Western civilization. Philip II of Spain sailed on the Protestant England of his sister-in-law Elizabeth I in order to make the world safe for Catholicism and trade without fear of pirates. (England's privateers figured why go all the way to the New World when you can just steal the stuff from the Spanish ships in the Atlantic.)

"If the Spanish Armada had landed," wrote Felipe Fernandez Armesto in *The New Statesman,* "English resistance would have crumbled. . . . Elizabeth I would have done a deal. . . . There would have been no Protestant England, no independent Netherlands, no seventeenth-century conflict of conscience, no English civil war, no United Kingdom, no British empire and above all, because there would have been no 'Pilgrim Fathers'—they would have been immolated by an English Inquisition—there would have been no United States of America." All of this might have been our history, were it not for the direction of the wind.

The Spanish Armada, assembled by Spain's Catholic king

Philip II, must have been a sight to behold as it set sail on May 9, 1588. It consisted of 130 ships of various sizes, all decorated with religious banners and flags bearing the Holy Cross. Sixty-five of the vessels were warships loaded with cannons and weaponry. They carried an invasion force of 19,295 soldiers and 8,050 officers and mariners. In addition, there were 180 priests and monks who performed daily mass on board and prepared to do the converting of the English populace once the battle was over.

The huge fleet was to sail north and rendezvous with thirty thousand Spanish troops based in the Netherlands. They would create a combined force of fifty thousand men. From the beginning, the weather refused to cooperate. As the Spanish embarked from Lisbon, with the inspiring words of Philip II fresh in their minds—"the principal aim of His Majesty is the service of God"—the wind blew them in the opposite direction. By the time the wind changed course, they were at Cape Saint Vincent in the extreme southwest of Portugal. Finally, they got turned around, backtracked, and made it to La Coruña in the north of Spain, but just as they were finally going to get beyond the Iberian Peninsula, a storm hit.

Some ships took refuge in the harbor, but others were less lucky—they lost control and were blown out to sea. This forced another delay as the fleet reassembled itself, the broken ships were repaired, and supplies soaked with seawater were restocked. After the false starts, quite a few of the Spanish sailors were thinking that the trip was not as much fun as it was supposed to be, and guards had to be posted to keep them from deserting.

So with their morale slightly dampened, the fleet took off once more, and after two months the Armada finally spotted the English coast. The English had also spotted the Armada. Part of the English fleet, which consisted of 175 ships, took up a position behind the Spanish. They managed to capture two Spanish ships that had both been damaged by accidents. One had been accidentally rammed by another ship; the other was nearly sunk when its magazine exploded. For more than a week, the two sides fought a series of small skirmishes, which kept the Spanish from meeting up with the army in the Netherlands, where they had hoped to replenish their stores of ammunition. No such luck. By July 28, they were running dangerously low. What is more, the winds were still working against them. The English referred to the breeze that favored them as a "Protestant wind."

This was when the English sent eight explosives-laden ships into the middle of the Spanish fleet and deliberately set them ablaze in order to create panic and force the Spanish to break formation. The fleet dispersed in confusion, racing for the open sea. In their attempted escape, several ships collided; one of the largest ran aground. A few stragglers, eleven ships, were left behind and surrounded by about a hundred English ships, which blasted away at the outnumbered and nearly defenseless Spanish all day. Amazingly, not a single ship sank until that night, when a strong wind capsized the *María Juan* and two others ran aground. On August 9 the wind changed direction again and transported the remainder of the Spanish Armada northward, out of the Channel. Here they had two choices: They could turn around, face the English, and battle on; or they could take advantage of the wind for a speedy passage to the North Sea. They decided to cut their losses and head home by sailing up around Scotland, then back down past the western coast of Ireland.

Yet the Spanish were unable to retreat from their battle with the elements. In the North Sea they were met by constant storms and freezing fog. As they rounded the Shetland Islands at the extreme north of the British Isles, it stormed for four nights, and seventeen of the ships disappeared in the fog. Most of those lost sank or ran aground off the Irish coast.

Captain Francisco De Cuellar of the *San Pedro,* which sank off the Irish coast, described the scene. "I placed myself on the top of the poop of my ship, after having commended myself to God and Our Lady, and from thence I gazed at the terrible spectacle. Many were drowning in the ships, others, casting themselves into the water, sank to the bottom without returning to the surface; others on rafts and barrels, and gentlemen on pieces of timber; others cried out aloud in the ships calling upon God; captains threw their jeweled chains and crow-pieces into the sea; the waves swept others away, washing them out of the ships."

Some waterlogged Spanish sailors managed to swim to shore, but most did not receive a warm welcome. The English in Dublin, afraid of what the Spanish could do in this poorly defended territory, gave orders that all Spaniards be killed. Only a few, Captain De Cuellar included, made it back to Spain with the help of the Irish Catholic underground. In all, about twenty-four Armada ships finished their journey on the coast of Ireland, and ten thousand soldiers lost their lives there. The bodies of Spanish sailors and the wreckage of their ships littered the Irish shore for miles. Of the proud Armada of 130 ships, only 80 ever returned to Spain.

For the English, however, this was all a cause for celebration. Queen Elizabeth even commissioned a special medal to commemorate the victory. It gave all the credit to the "Protestant wind."

Thanks to the Wind the Lost Colony Remains America's Greatest Mystery

On August 17, 1590, John White arrived on Roanoke Island after a long sea voyage from England. The trees were in full bloom and parakeets flew from branch to branch against a clear, sunny sky. White was about to be reunited with his daughter Elynore; her husband, Ananias Dare; and his granddaughter, Virginia Dare, the first child born to English parents on American soil. They were part of a colony of more than a hundred settlers of which White had been governor. White was an artist who had spent most his time in the colony drawing detailed maps and sketches of the native plants, animals, and people.

White's colony had not been the first European attempt to settle Roanoke Island, North Carolina. The first expedition, made up of all men, was designed to find an appropriate piece of real estate, leave fifteen men there, and then send the ships back to collect a new wave of colonists, including women and children.

The romantic portraits of a boundless new land of adventure and plenty painted by the boasting men who came back from a

New World adventure were great advertising. Arthur Barlowe, one of the captains of the original expedition, wrote that North Carolina was a delicate garden, full of fragrant flowers. The land was "the most plentiful, sweet, wholesome and fruitful of all the world." The Native Americans were "gentle, loving and faithful, void of all guile and treason." There was no shortage of volunteers to sail to this nirvana. It was the first time an American colony would include women and children. Only by settling entire families, the English believed, could Roanoke Island become a new England.

In fact, the native inhabitants of Roanoke Island were kind to the settlers at first, but after the departure of the English ship, the commander of the first colony, Sir Ralph Lane, had treated them so harshly that their relationship with these Europeans took an extreme turn. Even as Barlowe was attracting new adventurers to North Carolina with his tales of friendly natives in the American Eden, the Indians were attacking the foreigners. The fifteen members of the original colony scattered and were never seen again.

Meanwhile, a second set of colonists, including White and his family, sailed toward a new land and a bright new future blissfully unaware of what had transpired there. The plan was for the second group to pick up the original colonists and proceed to Chesapeake Bay, where they would set up a new town. But when they arrived at the location of the original colony to find nothing but abandoned, burned-out houses, they were too dispirited to get back on the ship. So instead of moving farther along the shore, they made their camp in the exact same location as their predecessors.

They spent a month tilling the soil, drawing water, and rebuilding the houses. After about a month, the families were comfortable enough in their new homes to send their ship back to England for supplies. When John White kissed his daughter and boarded the ship, he had no idea that it would be the last time he would see her. The whole trip was supposed to take only three months, but war broke out between England and Spain, and all available ships were put into war service. It would be more than three years before he could get back to his family and friends in America.

White's happy reunion was not to be. He stepped onto the sandy ground, sounded a signal trumpet, and waited—no response. He wandered into the site of the former settlement to find nothing but an abandoned fort, a few metal objects, and a post with the letters CRO carved into it. A few moments later, he saw a message carved into the bark of a second tree: "Croatoan."

It wasn't a lot to go on, but White was certain that he knew what it meant—the colonists must have gone to live on the island of Croatoan with the friendly Native Americans. He implored the captains of the expedition to sail to Croatoan to look for his family, but before they were able to chart a course, a huge wind severed the anchor of one of the ships and tossed the others around so fiercely that the captains were afraid they would be dashed to bits. They refused to spend another minute on the Carolina coast, and instead returned to port for refitting and wintering. The wind had ensured that the fate of the Roanoke Island settlers would always be a mystery. The stump of the oak on which the word Croatoan was carved remained the only evidence of the famous lost colony. It stood until 1778 as a historic monument.

What happened to the colonists? They, too, may have succumbed to the weather. By studying the rings on trees, scientists have discovered that from 1587 to 1589 the East Coast of America suffered an extreme drought, the driest that part of the continent had been in eight hundred years. When food became scarce, the unprepared colonists must either have starved or gotten into a war with neighboring tribes over resources. Some scholars believe that the colonists were captured by the Indians and sold into slavery;

others envision something much more peaceful. Perhaps the English moved inland among friendly Native Americans, intermarried, and dispersed.

A group of archaeologists known as the First Colony Foundation is now hoping to unravel the mystery. They started digging in Fort Raleigh Park, looking for artifacts and clues. One of the problems is that no one is exactly sure where the original colony was located. Over the years, vegetation, sand, and perhaps water have covered the footprints. Underwater archaeologists believe that as much as a quarter mile of the island may have sunk beneath the sea since the sixteenth century.

Yet some history buffs are grateful to the wind that kept John White from learning the fate of his kin. "I've always said I'd be just as happy if it was never solved," Phil Evans, a founder of the First Colony Foundation told *National Geographic*. "As long as the Lost Colony is unexplained, it stays fascinating for a lot of people . . . they learn history. I don't want to take away the mystery."

Gee, It's Cold in Russia, Part I
Charles XII Invades Russia

Should you ever get it in your head to invade Russia, be sure to pack extra supplies of long underwear and warm gloves. This is a lesson that has been learned at a great cost in human suffering, and it has been taught over and over to leaders who cast their covetous gaze on that huge northern expanse. Again and again, would-be conquerors of Russia found that if Russian weapons were not enough to kill you, the elements would. In 1709 the young Swedish king Charles XII became the first great European invader to lead his men on a long march of death and exhaustion through the Russian winter. His army's destruction by mud and cold would be repeated in later centuries by Napoleon and Adolf Hitler.

The winter of 1708–1709 was not made for outdoor camping. In the throes of the Little Ice Age, all of Europe was frozen solid. We have already discussed some of the effects of the era's climate change in Greenland and Germany. In normally temperate Venice, the canals iced over, the courts of justice in Paris were closed due to cold. You can imagine, then, what awaited the army of Charles XII as it marched into Russia in an attempt to end the Great Northern War.

The war began in 1700, when Peter I (he would later be known as "the Great," but for now was just Peter the so-so) declared war on Sweden in an effort to drive Sweden out of the Baltic region. Declaring war was a monumental act of hubris for a Russian czar. Sweden was a superpower. Its sphere of influence included modern-day Sweden, Finland, Estonia, and parts of Latvia and Russia. In fact, Sweden controlled the area where St. Petersburg now sits.

Russia's history up to that point was shaped more by its military defeats than its victories. The word "Russian" itself may refer to the red-haired Vikings who arrived in Kiev (now in Ukraine) in the eighth century. A Scandinavian named Rurik was invited to rule over some hot-tempered Slavs who apparently had a bit of trouble getting along with one another. By the end of the ninth century, foreigners began to apply the term "Rus" to both the Scandinavians and the Slavs. In 911 the officials of the land signed a treaty with Constantinople and the signatories were Injald, Farulf, Gunnar, Frithleif, Angantyr, Throand, Leithulf, Hrolf, Vermund, Harald, Karm, Karl, Fast, and Steinvith—good Russian names all—and proof that the upper class of the land was entirely Nordic. Eventually the Slavs and the Scandinavians mixed and mingled into one group of Russians. But Russia's record in battle after that was not stellar; the nation had been overrun and enslaved by the Mongols and others. There was little in its past to suggest that it could match the mighty Swedes.

From the Russian perspective, the war got off to a rather unpromising start. In Peter's absence, the Russian army had been thoroughly trounced in the battle of Narva by a vastly smaller Swedish force. The victory gave Charles XII a great boost of confidence—in retrospect, perhaps too great a boost. After Narva,

Charles stopped worrying about Russia for a while. He took the time to invade Denmark, Poland, Lithuania, and Saxony before finally setting his sights on Peter's territory a full eight years later. Peter used the break in the action to build up his military forces—and for good measure, to build the city of St. Petersburg.

Even so, the Swedes were the initial victors in the battle of Holowczyn in July 1708. From there Charles planned to march on Moscow, but Peter surprised the Swedes with a completely unexpected tactic. As the Russians withdrew, they set fire to their own land—they burned all the houses, crops, and tools in their path so that the advancing enemy would find nothing of value. Without shelter and with the winter rolling in, the Swedes were forced to change their course, and they marched south toward Ukraine. Ukraine was rich in fruit, cereals, and grazing livestock, and it would have been an ideal place for the Swedish army, had they appeared two months earlier. But they arrived in November, as one of the coldest winters in memory was about to set in. The Swedes, of course, were no strangers to cold weather, but even the heartiest soldier could do battle only so long against the elements huddled in the rare shack that the Russians had failed to burn.

"Fighting was but child's play compared to what we endured," wrote Carl Klingspor, a soldier in Charles XII's army who survived many a battle without injury only to lose two fingers and an ear to frostbite. "Around us, amid the howling icy blasts, the very game froze stiff in the field and the birds fell stark dead from the air, even as if they had been shot. . . . Our hearts ached and our eyes ran when we saw the hundreds of brave lads who cried for the field-surgeons to cut off their hands and feet that had grown white and crackly, while the ears and tips of the noses would drop off without even the assistance of the knife."

In one march alone, two thousand men fell dead from exhaustion and frostbite. The living were in pain, unable to use their hands, frozen to their mounts. Some would die when they sat down by the fire, as the rapid heating caused a sudden flow of blood into hopelessly constricted veins. As the winter wore on, it was hard to recognize Charles's men. Not only were they often disfigured from their bout with the elements, but they were dressed in the uniforms of felled Russian soldiers.

"If we went out hunting for them," Klingspor wrote, "it was not now for the pleasure of their killing, but even as we would hunt certain game, for the warmth of the coat we might thereby obtain."

The severe cold also took its toll on the arsenal. First, the cattle and horses began to fall dead. Without the cattle, there was no way to pull the cannon. The gunpowder was soaked with rain and snow, and Swedish shots left their weapons with a dull thud and little power.

The Swedish forces, once fifty-one thousand strong, had dwindled to twenty thousand by spring. Of those remaining, more than a third were ill or maimed. It was this winter-weary force that would meet Peter's battalions at the battle of Poltava.

Despite their winter attrition, the Swedes once again took the offensive. In early January they attacked the little fortress of Veprik, which was taken easily, but at the cost of a thousand Swedish lives. After a couple more engagements, the Swedes encountered another facet of the Russian weather—*rasputitsa*. In the spring, all the ice and snow melts too quickly for the ground to absorb it. The resulting mucky quagmire makes a mockery of cart wheels. There was no choice but to sit and wait for the soil to solidify.

Charles set upon the city of Poltava in May 1709. The Russians set up entrenchments within a few hundred yards of the Swedish lines in order to force the Swedes to attack. When Charles learned on July 7 that forty thousand Russian reinforcements were to arrive in two days, he made the fateful decision to strike immediately. Charles, himself injured in battle, planned to charge past the Russian line and straight into the main Russian defensive position. But the Swedes, who had endured so much over the winter, now numbered only seventeen thousand. They were no match for a counterattack by forty thousand fresh Russian troops. Nearly the entire Swedish army, aside from Charles and fifteen hundred of his closest followers, were killed in the battle. The survivors escaped to Turkish territory.

This battle was not the last of the Great Northern War, which would drag on for another twelve years. It was, however, a major turning point. It allowed Peter to build up the great Russian naval strength in the Baltic Sea that would finally bring Sweden to its knees. Even more important, however, was the psychological impact. Word of Charles's defeat resounded throughout Europe. Sweden's status as a major power would soon come to an end, and the world was on notice that Russia and Czar Peter I were a force to be reckoned with.

As Lt. Col. Joseph B. Mitchell wrote in his update of Sir Edward Creasy's classic *Twenty Decisive Battles of the World*: "With the downfall of Sweden there disappeared from the scene the only power in the Baltic that could have opposed the growing strength of Russia. The decisive battle at Poltava was therefore all-important to the world, because of what it overthrew as well as what it established."

The world should have noted, as well, that Russian climate and geography are forces to be reckoned with—but later invaders failed to heed this lesson.

The Secret of the Stradivarius

Among musical instruments, the Stradivarius has no peer. Violins crafted by the master Antonio Stradivari are renowned for their rich, powerful sound. The master produced at least 1,116 instruments, of which 540 violins, 12 violas, and 50 cellos survive today. So prized are they that a single violin has been known to fetch as much as $4 million. For years, musicians and scientists have sought to unravel the secrets of Stradivari's method. Just what is it that makes a Stradivarius so resonant? There have been many theories: some believe it was the formula for the varnish; others say it was a secret Italian manufacturing method. Modern climatologists have come up with a new explanation: it may have been the weather.

Cremona, Italy, of the sixteenth century is the birthplace of the violin. Of course, stringed instruments were nothing new—back in the eighth or ninth century Asians played stringed instruments with bows. Somehow a few made their way to Europe, where they evolved into a vielle, a thirteenth-century creation that was similar to a violin but flatter, which in turn evolved into the German *klein geigen*, or little fiddle.

Then came Andrea Amati. Born about 1510, Amati is considered to be the inventor of the modern violin. His grandson, Nicolò, improved upon Grandpa's design. Nicolò, in turn, trained Antonio Stradivari, who started producing violins under his own name while still an apprentice.

At first, Stradivari followed the example of his mentor. His violins were small and solidly constructed, with a thick yellow varnish. But he gradually cultivated his own style, and in 1684 started producing larger models with a deeper-colored varnish. The proportions were innovative, and so was the formula for the varnish.

Stradivari mixed silica and potash and applied the result to the wood. It soaked into the pores and wrapped around the fibers. The mixture has done a wonderful job in preserving the wood of the instruments to this day. In the second phase, the master applied an insulating coat made of egg whites and honey. This gave the violin a glossy finish (and presumably made it a great dessert). Then came the final varnish, the exact composition of which has been lost, but it contained some mixture of dyeing substances like propolis, gum Arabic, turpentine, and a resin. The secret mixture was long thought to be the secret to the Stradivarius's sound. But most researchers agree that it was not merely the varnish, but the combination of craftsmanship, varnish, and—perhaps most important—the right wood.

No musical instrument produces a pure tone. Each note is accompanied by a series of overtones, which are multiples of the basic pitch and frequency. The number and volume of the overtones are what make a piano, tuba, and violin sound so different even when they are playing the same note. The properties of wood are integral to creating these overtones. There is the elasticity along and across the grain, the damping characteristics of the wood, the density and velocity of sound through the wood. All of these affect the way a violin reacts to the vibrations of the strings.

If you don't think the grade of wood matters, imagine the sound of a plywood violin.

Two researchers, Henri Grissino-Mayer, a University of Tennessee tree ring scientist, and Lloyd Burckle, a Columbia University climatologist, studied Stradivari's methods and published their conclusions in the journal *Dendrochronologia*. (That's the scientific study of tree rings to you and me.) They believe it was the sharp dip in temperatures of the Little Ice Age that created the wood that made the violin that so many revere.

The coldest snap of the Little Ice Age was the Maunder Minimum, named for the astronomer E. W. Maunder, who documented a lack of solar activity during the period. The Maunder Minimum started in 1645, one year after Stradivari was born, and continued until 1715. This cooling trend affected the rates of tree growth. Trees growing during this period showed the slowest growth rates of the past five hundred years. The wood, with narrow growth rings, was especially dense and strong, the perfect material for a master craftsman. Stradivari used local spruce wood in the manufacture of his violins, which were created between 1666 and 1737, with the most prized and valued violins created between 1700 and 1720.

"The onset of the Maunder Minimum at a time when the skills of the Cremonese violinmakers reached their zenith perhaps made the difference in the violin's tone and brilliance," wrote the scientists.

The conditions that created the violins of Stradivari and his peers, that particular combination of craftsmanship, attention to detail, and climate, are not likely to be repeated anytime soon. Before you shell out a few million for one of the master's works, however, you should know that from the concert stage most listeners can't tell the difference. A modern violin can produce as beautiful a sound in the hands of a skilled musician. The world's most prized Stradivarius will produce nothing but noise in your eight-year-old nephew's Suzuki class.

Another Protestant Wind Blows a New King to England's Throne

In 1688 the people of London were anxiously eyeing a dragon, desperate to see if its head was where its tail should be. The dragon in question was a weathervane atop the Church of Saint Mary le-Bow. The mythological creature was the symbol of London, and all eyes were on it to see which way the wind was blowing. Everyone knew that the Protestant William of Orange would soon sail from the Netherlands for England, ruled by the Catholic James II. The success of his invasion would depend upon favorable winds, and those meant winds from the east—a "Protestant wind," not a westerly "popish wind."

In the good old days of 1688, international foreign policy tended to consist of family spats on a grand scale (this, in fact, remained true through World War I, the last major entanglement between blood-related regents). Most of the rulers and nobles of Europe were part of one royal family tree, which looked a bit more like a tangled shrubbery.

James II was the son of King Charles I (a Protestant) and Henrietta Maria (the Roman Catholic sister of the future Louis XIII of

France) and brother of King Charles II, who had no children. Charles I, who advocated the divine right of kings, and who enraged the Scots by trying to impose Anglicanism on them, was impeached and sentenced to death in 1649. England survived without a monarchy for a little more than a decade, but it was restored in 1660, with the younger Charles assuming the throne. James converted to Catholicism in the late 1660s and married the Catholic Mary Beatrice of Medena. Shortly thereafter, Parliament passed a law with only one man (James) in mind—it forbade Roman Catholics from holding public office. The House of Commons later made an exception for the very man for whom the law had been created in the first place. James came to believe that the reason his father, Charles I, had been unpopular was that he was not firm enough. James was nothing if not firm in his speedy moves to appoint Catholics to important government positions. He created a large army and imported Catholic soldiers from Ireland to make sure the Protestant majority knew which way the metaphorical wind was blowing.

William of Orange had an extremely tenuous claim to England's throne, but since when did that stop anybody? He was born to William II of Orange and Mary Stuart, who was King James II's sister. The younger William was also married to a Mary Stuart (aka Mary of York); this Mary was the Protestant daughter of the current King James II and his first wife. So William of Orange was James II's son-in-law and his nephew.

In 1687 James had no sons and no children from his second marriage. If he were to die the throne would go to William's wife, Mary, a devout Anglican. When word spread that Mary Beatrice was in the family way, a sense of urgency was added to the Protestant cause. If the child of James and Mary was a boy, by virtue of his Y chromosome, he would succeed to the throne. In June their fears were realized when James Francis was born.

The Dutch Republic had political and economic reasons to intervene in England's affairs as well. Louis XIV of France had been imposing heavy customs duties and prohibiting the import of certain Dutch products. It was a serious blow to the economy of the shipping giant. (The Netherlands had an estimated fifteen thousand merchant ships compared to about five thousand for the rest of Europe.) If an economic and military alliance was formed between England and France, it could spell trouble for the Netherlands.

Back in England, there was growing dissent among the Protestant community. Rumors ran rampant that the new baby was not James's son at all, but a servant child smuggled in by Jesuits. An increasing number of Protestants were praying for the arrival of William of Orange. They watched the dragon and waited. But the wind, in the words of the British historian T. B. Macaulay, "blew obstinately from the west, and which at once prevented the Prince's armament from sailing and brought fresh Irish regiments from Dublin to Chester, were bitterly cursed and reviled by the common people. The weather, it was said, is Popish. Crowds stood in Cheapside gazing intently at the weathercock on the graceful steeple of Bow Church, and praying for a Protestant wind." (Cheapside, incidentally, referred to the main market area of London. It took its name from the Old English *ceap*, which once meant *market* but has evolved into our modern word *cheap*.)

As they waited, many of the Protestants took to singing a song called *Lilliburlero*, which had been written in 1687 when another "Protestant wind" kept the Irish Catholic Earl of Tyrconnel from proceeding to Ireland. It said: "Oh, but why does he stay behind, Ho, by my shoul, 'tis a Protestant wind."

In late September William's men began boarding their ships. There were around forty-five vessels (historical accounts differ on the exact number) stocked with cannons, soldiers, and thousands of horses. They hoped to set sail before the cold weather blew in, but as September became October, the wind continued to blow from the west. William waited. For the next eighteen days or so the wind was both popish and stormy. "One night the Winds were so very high, and the Air so tempestuous and stormy shaking the very Houses and People in their Beds, that the whole Fleet riding at anchor was in great peril," wrote one witness.

For most of late October, agnostic winds from the northwest blew—then on October 24 or so, they swung to the southeast and finally from the east. On October 30, William and his men finally ventured into the sea, only to encounter another storm. The skies turned dark, the ships rocked, and the men all got seasick. What is more, the horses had not been tied properly and some were rearing and running about, creating chaos. Although no ships were lost, many were damaged and four hundred horses jumped overboard. The story of the damage to the fleet reached England through word-of-mouth and the destruction to the fleet became greater and greater with each retelling. Priests in England were boasting about how their prayers had been answered, and James was now certain that an attack was not forthcoming. Since his position no longer seemed at risk, he rescinded a promise to hold parliamentary elections in November. He may have been safe from William for the time being, but his turnabout won him few friends at home.

Meanwhile, in the Netherlands, William's navy was repairing its ships, and the cavalry was restocked with horses. The Dutch did not see the delay as a great problem, since stormy weather would have kept the fleet in the harbor at any rate. The conditions finally improved on November 11, but they had long missed their chance of sailing before the cold weather set in. As they traveled, they were pelted with sleet, but William sailed on. On November 13 the armada passed the English navy—the English wanted to give chase, but the floodtide was against them. The Dutch simply sailed by as crowds gathered on the shore to watch them. This is when the Protestant wind began to blow and helped the Dutch sail westward into the English Channel. They landed in Torbay, in southwestern England. They were a bit surprised to find there were no soldiers in Torbay to oppose them. The English navy had started its belated pursuit, but because of the wind they could make it only as far as Portsmouth. On November 17, when the English ships tried to enter the Channel, the winds were so strong from the southwest that they could not enter and had to take shelter elsewhere.

James, with his nerves shattered and suffering from persistent nosebleeds, took refuge with his wife and son in the court of Louis XIV. Less than two months later, William and Mary, Prince and Princess of Orange, were named king and queen of England. William, by virtue of the Y chromosome, was granted sole administrative authority.

Ben Franklin and That Kite

Every American schoolchild knows the story. Old Benjamin Franklin discovered electricity by flying a kite, with a key tied to the string, during a thunderstorm. As they get a bit older, most schoolkids start to wonder, "Does this make any sense?" Just how did Ben manage to perform this kite trick without getting fried in the process? Then it's time for recess, and the whole matter is forgotten. For those who still occasionally wonder, I have the answer. It turns out that the kite incident may not have happened at all.

This is not to say that Benjamin Franklin was not a pioneer in the electrical field (as it were). In 1746 he was quite wealthy as a printer and the publisher of *Poor Richard's Almanack*, which frequently poked fun at the astrological and meteorological superstitions advanced by the other almanacs of the period. One day, Franklin saw a parlor trick called "the electric kiss" by a Scottish showman, Dr. Adam Spencer, aka "Dr. Spence." In his show, a woman would put her bare hand against a spinning glass globe, and as her suitor came in close for a kiss, sparks—and merriment—would fly. Electricity became Franklin's passion—he just had to tinker with Dr. Spence's Leyden jar. A Leyden jar is a primitive version of what we would now call a capacitor. First created by Dutch scientist Pieter van Musschenbroek, a professor at Leiden University (hence Leyden jar), the device was made of an ordinary jam jar coated on the inside and on the bottom with tinfoil. A waxed cork was fitted to the mouth of the jar and a brass rod ending in a short chain was pushed through the cork. The chain made contact with the interior foil and the brass rod had a brass knob on top. When a static charge is created by rubbing various objects together and is applied to the external brass knob, the static charge could be stored in the jar by way of the brass knob. Because unlike charges attract, a negative charge on the inner plate would induce an equal, positive charge on the outer plate. Franklin bought the good doctor's entire setup and asked a friend in England to send him some more equipment and articles on experiments. Franklin's electrical hobby was soon his full-time occupation.

The best thinkers in Europe at the time claimed that there were two types of electricity. Static electricity produced by rubbing a glass rod would attract a ball, but electricity produced by rubbing a resinous rod repelled the same ball. This was evidence that there were two different varieties. The amateur scientist from the colonies begged to differ. He believed that electricity flowed from greater charge to lesser charge. He identified and named positive and negative charges, conductors, and insulators. His most controversial claim was that lightning was an electrical phenomenon. Franklin corresponded with Peter Collinson, a Fellow of the Royal Society, about all of this, and the letters formed the basis of a book, *Experiments and Observations on Electricity,* which was first published in 1751. The book was translated into several languages and earned the printer a worldwide reputation as a natural scientist.

The book proposed an experiment to test whether lightning and electricity were, in fact, the same thing. "To determine the question of whether the clouds are electrified," he wrote, "on the top of a high tower, place a sentry box big enough to hold a man and an electrical stand. From the middle of the stand let an iron rod rise and pass bending out of the door, and then upright twenty or thirty feet, pointed very sharp at the end. If the electrical stand be kept clean and dry, a man standing on it when such clouds are passing low, might be electrified and afford sparks, the rod drawing fire to him from a cloud."

The most noted electrical experimenter in Europe at the time was the Abbé Nollet in France. He gave no credence to the ramblings of the colonial scholar. King Louis XV, au contraire, was intrigued, and he encouraged his scientists to try the experiment and confirm the theory. On May 10, 1752, Thomas François d'Alibard rose to the challenge. He built a pointed metal rod forty feet high in the gardens at Marly-la-Ville, France, and waited for a storm. As the storm clouds flew over, an assistant moved a brass wire toward the contraption and attracted sparks, just as Franklin had predicted. French scientists repeated the experiment throughout the summer. It was confirmed—lightning was an electric spark. Almost overnight Benjamin Franklin was a celebrity in Europe, and not a single kite had been flown.

Benjamin Franklin never wrote up a kite experiment. There is no journal recording such an event. The first mention of a kite experiment was a letter to the *Philadelphia Gazette* dated October 19, 1752. The letter outlines an experiment but never says that Franklin actually carried it out.

"Make a small cross of two light sticks; to reach the four corners of a handkerchief. To the top of the stick attach a very sharp pointed wire . . . to the end of the line tie silk ribbon and fasten a key. The person who holds the string must stand within a door so that he does not become wet. As soon as the thunderclouds come over the kite, the pointed wire will draw the electrical fire."

Note that in the experiment, the lightning did not directly hit the kite. The idea was for the kite to travel close to the clouds to draw a charge from the weather system down the string. Then the experimenter would bring his knuckle to the key and drain the charges off into the ground. Perhaps Franklin tried this himself, but if he did, he left no firsthand account. The most detailed account of Franklin's flying a kite comes from Joseph Priestley, who claimed Franklin told him about it. Priestley's secondhand account of the experiment was published in 1767.

At least one person, the Russian physicist G. W. Richman, was killed trying to repeat the experiment. Richman flew his kite in St. Petersburg in 1753 and was struck down by "a palish blue ball of fire, as big as a fist" that "came out of the rod." More recently, Tom Tucker, a lecturer and historian at the Isothermal Technical College in North Carolina, sought to repeat Franklin's kite experiment. In 2003 he tried to re-create it using materials that would have been available in Franklin's time. He followed the description exactly and tried it several times, but it wouldn't fly. He tried again using a modern kite, but it didn't work, either. He now believes that it could not have worked for Benjamin Franklin.

Those in the Franklin-did-fly-the-kite camp have a number of theories as to why he never gave details about his experiment—where it happened and when. Perhaps he was embarrassed to admit playing with a child's toy. In any case, experimenting with kites was not Franklin's invention. Three years earlier, a physicist named Alexander Wilson had raised a train of kites to do weather research. The U.S. Weather Bureau started using kites for research on a regular basis in 1893. As balloons, airplanes, and satellites offered new ways to gather information, kites became toys again. The last kite weather bureau closed in 1933.

Whether or not he ever flew a kite into a storm, Franklin was quite correct about lightning rods. He was elected to the Royal Academy of Sciences in 1756 and received the prestigious Sir Godfrey Copley Medal. The great cathedrals of Europe would soon be fit with rods. The Campanile bell tower in Venice had been struck by lightning and burned six times since 1388, but after "Franklin rods" were installed, they had no more trouble. You would think the protection of steeples would make Franklin particularly popular with the clergy, but this was not always the case. "Benjamin Franklin's lightning conductor is a sacrilege that tries to avert the wrath of God," said a Boston minister in 1753. "The destruction of Lisbon by the earthquake and tidal wave is God's punishment of man for his sacrilege."

No matter. Benjamin Franklin was the toast of the scientific community, and his popularity in Europe, particularly in France, would prove to be a great asset to America when it sought to win its independence. His reputation and close relationship with King Louis made him the perfect ambassador to plead for French aid. Thus Ben Franklin's fascination with lightning was important in convincing the French to side with the revolutionaries.

In England, King George III's distaste for Franklin's politics spilled over into the scientific realm. He ordered that no Franklin rods be put on the palace. He suspected that they were some kind of trick. Instead of transporting the electricity harmlessly into the ground, they might be a sneaky way for rebels to channel lightning into royal buildings. He commanded the president of the Royal Society, John Pringle, to disclaim Franklin's theories of electricity. Pringle immediately resigned. "You may alter the laws of the land," he said, "but not the laws of nature."

Through Many Dangers, Toils, and Snares

"Amazing Grace" is one of our best-loved and best-known hymns. The myth of its creation goes something like this: John Newton, the captain of a slave ship, was transporting his human cargo across the Atlantic when a huge storm nearly tore his vessel apart. In the midst of the storm, the captain made a promise to God that if he survived he would devote his life to His service. When he made it through the storm unscathed, he saw the error of his ways, turned the ship around, and wrote the song that we know to this day. Like all myths, this one is based on a kernel of truth, but through repeated retellings it is much improved. The facts are these: the storm was real; the slave ship was real (although Newton was not the captain); and Newton would, much later, become a minister and write "Amazing Grace," among his many recitations (he did not write the music). He later became an abolitionist.

Slavery is almost as old as human civilization. There were slaves in ancient Mesopotamia, Greece, and Rome. The Bible mentions slaves and the medieval church condoned slavery, except when

Christians were enslaved by infidel Muslims during the Crusades. The fifth-century Anglo-Saxon word for slave was *Welshman*, which shows what ethnic group they preferred to enslave. The modern English word *slave* is derived from the Slavic people the Germans captured and sold in European markets. The Portuguese and Spanish tried to make slaves of the indigenous Americans. Unfortunately for the Indians, they were not resistant to Spanish germs. When a plague cut through the native population, the Europeans had to look elsewhere.

Interestingly, two future presidents were indentured servants as youths. Indentured servants were under contract to work for another person, generally without any pay, in exchange for something of value like a free passage to a new country. Until the end of the contract they were not at liberty to leave or to refuse the contract holder's demands. But unlike the African slaves, such servants were free to leave after a fixed number of years. Still, the rights of an individual servant were in many ways comparable to the rights of a slave. Millard Fillmore and Andrew Johnson were both bound by such contracts. Andrew Johnson escaped, and his master placed an advertisement in the Raleigh, North Carolina, *Gazette* offering a ten-dollar reward for his capture and return. Fortunately for history, the future president was not caught. Fillmore was more patient and purchased his freedom for thirty dollars after serving his master for several years. This type of servitude, however, bore little resemblance to the African slave trade at its height, when the American colonies created a mass market with a seemingly inexhaustible need for new, free labor.

Slavery was such an accepted form of commerce in the early to mid-1700s that few stopped to question it. Slave merchants were devout churchgoers and pillars of the community. Some Christians argued that slavery was God's way of rescuing Africans from their barbaric, heathen lives and introducing them to the true religion.

What is more, they argued, it was good for Africa. As the demand for slaves grew, a powerful merchant class was created in countries like Nigeria. The slave trade created new settlements and removed socially undesirable elements from African communities. It is true that the slave trade *was* a boon to certain elements of African society. Often, mixed-race children, the product of unions between slave traders and local women, held posts of high status. "Many Nigerian middlemen began to depend totally on the slave trade and neglected every other business and occupation," wrote Michael Omolewa in his *Certificate History of Nigeria*. "The result was that when the trade was abolished those Nigerians began to protest. As years went by and the trade collapsed, such Nigerians lost their sources of income and became impoverished." Only the Quakers and Anabaptists, as a matter of faith, condemned the practice.

Still, an assignment on a slave ship was not a coveted job. It was dangerous, unpleasant work. Even those who had no moral qualms about the job could not help but find the conditions in the cargo holds fairly distasteful. Alexander Falconbridge, a surgeon on African slave ships who later became an abolitionist described it this way: "They are frequently stored so close as to admit of no other posture than lying on their sides. Neither will the height between decks, unless directly under the grating, permit them the indulgence of an erect posture."

The risk of disease on ships packed with human beings was great—dysentery was always a problem. Then there was the possibility of violence. The slaves outnumbered their captors, and if they were to rise up, the sailors could well be slaughtered. In addition, slave ships were not what you would call the finest of the fleet. With the dramatic climate changes between Africa and Europe, they rotted quickly and usually had to be scrapped after half a dozen voyages. The merchants who owned them were notorious for cutting corners to reduce costs, which also reduced the insurance premiums on the ships. They were disposable vessels, made with the cheapest materials, packed to the very limits of their capacity.

John Newton certainly never intended to be a slaver. It was just a bit of bad luck on his part. At the age of seventeen, he'd been given a great opportunity to travel to the Caribbean and learn the plantation business. As he was in London, waiting for his ship to sail, he fell head-over-heels in love with a thirteen-year-old named Mary Catlett. This overrode his sense of responsibility, and he missed his ship. His father arranged for him to join the crew of a Mediterranean merchant ship, but again he somehow forgot to get on board. He was eventually pressed into service in the Royal Navy, but his record did not improve. He jumped ship to go see Mary, and found himself flogged and stripped of his rank. This is how he ended up on a slave vessel bound for Sierra Leone.

Long voyages at sea were generally not exciting. Samuel Johnson described sailing as "being in jail, with the chance of being drowned." So the restless youth passed his time making up songs. They bore little resemblance to the hymns for which he would later be known. These songs were satirical ballads much more popular with the crew than with the captain. Bawdy tales, foul language, and drinking were common on board, as was recreation involving female slaves. Among select sailors there was similar recreation involving male slaves, and by some accounts, with sheep. When they arrived on African shores, the Europeans felt free of their everyday moral codes—word of their behavior would never reach home. Newton most likely had some temporary African "wives."

After nearly two years in Africa, Newton joined the crew of a ship called the *Greyhound*. Over the next year, she traveled along the African coast collecting gold, ivory, and wood, but not slaves. Most of the voyage was boring and uneventful. Newton passed the time hosting drinking contests in which sailors swilled gin from large seashells and took bets on who would fall down first. When this game lost its charm, Newton looked for anything available to read. He must have been very bored indeed when he picked up a copy of an instruction book for monks, *Imitation of Christ* by Thomas à Kempis. He fell asleep one night reading the fifteenth-century monk's musings on grace and salvation.

He awoke suddenly to find that the entire ship was pitching violently in a storm. Men above were running to and fro trying to keep the vessel from falling to pieces. Waves were washing over the deck, and the upper timbers on one side of the ship were gone completely. Newton started to climb a ladder for the deck, but another crew member asked him to go back and get a knife. He turned back, and another sailor started up the ladder. Just then a huge wave swept over the ladder, and the other sailor was sent to a watery death. Newton did not have time just then to reflect on his fortune. He joined the others nailing bedding and clothing over the broken timbers to try to keep the sea outside from coming in. He had to rope himself to a beam to keep from being washed away. Over the next twelve hours the storm rocked the ship, tossing it left and right, and up and down as though it were repeatedly falling from a cliff. Pigs, sheep, and poultry ran across the decks and then were washed away. The barrels of rations were smashed open and littered the sea. Newton, who had once sworn off Christianity as mere superstition, found himself praying to Jesus Christ for mercy.

After his brush with death, Newton was a changed man. When the battered ship finally reached Ireland, he sought out the first church and began a lifetime habit of praying twice a day; but this conversion did not cause him to renounce slavery. In fact, Newton finally married his beloved Mary and became captain of his own slave vessel *after* the storm. His logs contained references to cargo that were typical of the age. "Thursday, 13th June. . . . This morning buryed a woman slave (No. 47). Know not what to say she died of for she has not been properly alive since she first came on board." He also wrote of slave uprisings on board, "I hope (by Divine Assistance) we are fully able to overaw them now."

Yet, increasingly the captain was drawn to religion. He regaled whoever would listen with tales of what his near-death experience had taught him. In 1754 he finally gave up the sailing life, and ten years later he was ordained a minister of the Church of England. Around the same time he published a book, *Authentic Narrative*, which recounted the storm story. He was such a popular preacher that his church had to be extended to accommodate everyone who wanted to hear him. The preacher wrote his own chants. These were spiritual poems that were spoken, not sung. He wrote at least one for each week's service and two or three for the New Year's sermon. "Amazing Grace" was one of these compositions—one of nearly 350. It did not stand out to its author as particularly special. He never talked about writing the hymn and never wrote about it in his letters.

It was not until 1780 that he moved to London and met a group of ministers who opposed the slave trade. The abolitionist movement was gaining force in England at this time, but it would be another five years or more before Newton took a public stand on the issue. In *Thoughts Upon the African Slave Trade*, published in 1788, he wrote: "I hope it will always be a subject of humiliating reflection to me, that I was once an active instrument in a business at which my heart now shudders."

Appropriately enough, "Amazing Grace" may have been given its music by slaves. Thousands of American slaves were converted to Christianity. The plantation owners saw religion as a way to civilize the Africans. For the slaves, Christianity had a greater appeal—its message of a heavenly reward after earthly suffering had special significance. Since few slaves could read and write, they learned religious poems by rote and passed them on by reciting them in call-and-response fashion or by setting them to music. Some musicologists believe the tune to "Amazing Grace" is Scottish or Irish. It is, after all, one of only two songs that sound good on the bagpipes (the other is "Scotland the Brave"). Yet many believe that the melody was a folk tune from the plantations of the American South, possibly brought there from Africa. What everyone can agree on is that "Amazing Grace" has been recorded more than any other hymn in the world.

Washington and the Weather

What we now know as the American Revolution or the War for Independence could well have been recorded in history as nothing more than a footnote; a minor bit of unrest in the history of the western arm of the British Empire. The British certainly had advantages over George Washington's rebels. When Washington became commander of the American army, it consisted of volunteers without uniforms and often without weapons. The army, such as it was, numbered about twenty thousand men. The British army, on the other hand, was a well trained fighting force of about forty-two thousand with another thirty thousand German mercenaries at their disposal. Plus, the American forces did not have the support of all the colonists, about a third of whom continued to support the British government in what could have been remembered as a civil war. But the plucky farmer-soldiers of America did have a few things going for them. They had the home field advantage, they benefited from the ancient enmity between Britain and France, and they were lucky enough to have the weather on their side when it counted the most.

It all began when the British general Thomas Gage sent a force from Boston to destroy the American "arsenal" at Concord, Massachusetts. The British forces were met by a group of musket-wielding farmers at Lexington. Someone fired what history texts now call "the shot heard round the world," and England and the colonial army were at war.

The rebels' stronghold was Boston, which had long been a center of anti-British protest. There was fighting around the city throughout the winter of 1775–1776. General Washington moved his artillery to a position just south of Boston, where they could fire on British troops in the harbor and in the city. General Sir William Howe planned to attack Washington's men with a squadron of ships, but during the evening of March 5, a storm with gale force winds blasted Boston Harbor. It was, in the words of the Reverend William Gordon, "such a storm as scarce anyone remembered to have heard."

Howe was forced to call off his attack because of "the badness of the weather." The aborted attack was one of the final straws. Boston was such an opposition stronghold that it did not seem worthwhile to try to attack again. Instead, on March 17, 1776, Howe sailed away. The date is now a Boston holiday known as "Evacuation Day."

On July 2, 1776, the Continental Congress met and agreed to declare its independence from England. It was a date that John Adams believed would go down in history. "The second day of July 1776 will be the most memorable Epocha in the History of America," he wrote. "I am apt to believe that it will be celebrated by succeeding generations as a great anniversary festival." The declaration was posted two days later.

If the British had been at all complacent about this uprising in the colonies, they were not any longer. They assembled the greatest expeditionary force yet sent out from England. The thirty-two thousand men were led by General Sir William Howe. The force arrived in five hundred ships and established a base on Staten Island. Their plan was to isolate New York, and then the rest of rebellious New England, from the loyal colonies and bring the land back to order.

George Washington answered by sending one-third of his troops, about twenty thousand men, to Long Island. On August 22 the British landed about fifteen thousand men on the western tip of Long Island and made camp. A week later they attacked using a flanking movement on the American left, which had no natural anchor. Unfortunately for the British, a steady north wind and an ebbing tide kept the warships from moving into New York Bay and into the East River, where they could have cut off the only route of escape for Washington's men. Had General Washington been completely defeated at that point, the revolution might have been nipped in the bud, but then a storm rolled in with thunder, torrents of rain, and thick fog.

The British were content to wait it out—after all, they were warm and dry in their tents and ships. The Americans were out in the muck, without tents or sleeping bags. The American fort and trenches were completely flooded with men standing up to their waists in water in some places. It looked like the colonial army was doomed to certain defeat, and General Washington was forced to give the order to retreat across the East River. Unfortunately, the winds made the crossing too treacherous until about eleven o'clock in the evening on August 29.

The British could easily have attacked the retreating forces. Such a blow to General Washington's forces at the beginning of the war might have thwarted the revolution and dramatically changed North American history. They were saved from this fate by a thick fog, which blanketed Long Island beginning at two in the morning.

The Reverend William Gordon, writing in 1780, said: "Had it not been for the providential shifting of the wind, not more than half the army could possibly have crossed, and the remainder, with a number of general officers, and all the heavy ordinance at least, must inevitably have fallen into the enemy's hand. Had it not been for that heavenly messenger, the fog, to cover the first desertion of the lines, and the several proceedings of the Americans after day-break, they must have sustained considerable losses."

The Battle of Long Island was a great loss to Washington, with about twelve hundred Americans taken prisoner and four hundred killed, but thanks to the "heavenly messenger" of the fog, the rest of Washington's army escaped to fight another day.

A great turning point in the war was the Battle of Saratoga in 1777—British general John Burgoyne had failed to break through the American lines and open up a route to Albany, New York, and expected relief from General Henry Clinton in the south never came. Burgoyne was forced to retreat toward the friendlier environs of Canada. As he sounded the retreat, it was warm and clear; but a southerly wind was blowing. It began to rain and did not stop. The road became a bog and walking was a mucky endurance test. The wagons, buried to their axles, were abandoned with the tents and baggage on them. The mud slowed the pace of the march to less than a mile an hour. It was slow enough that the Americans, led by General John Stark, were able to cross the Hudson north of the British and cut off their retreat. On October 13, Burgoyne surrendered. The state of communications being what it was at the time, General Clinton detached seventeen hundred men to support Burgoyne's regiment on October 15—two days too late.

The American victory at the Battle of Saratoga in 1777 deeply wounded British morale. More important, it signaled to the rest of the world that England might be beaten. France and England were still ever-ready for a rematch of the Hundred Years War, and France joined the fray on the colonists' side. They also supplied troops and about 90 percent of the munitions used by the colonists. Spain and Holland would later side with the Americans as well.

In the final decisive battle of the war, the Battle of Yorktown, the British were again thwarted in their attempt to retreat, in part by the weather. England's hold on the colonies was eroding by 1781. General Lord Charles Cornwallis, the British commander in the southern colonies, found himself face-to-face with an American regiment under the leadership of the Marquis de Lafayette. In order to maintain his lines of communication with the main British army, Cornwallis was forced to retreat and retreat again until he found himself at Yorktown, Pennsylvania. General Washington ordered Lafayette to block Cornwallis's land route from the city. Meanwhile, Washington's twenty-five hundred soldiers and four thousand French troops under the Comte de Rochambeau cut off the route to New York. The main Franco-American force marched south to the head of Chesapeake Bay. A French fleet of twenty-four ships under Comte de Grasse blocked the sea route, leaving Cornwallis surrounded. His reinforcements never arrived.

On the night of October 16, 1781, the general made his last attempt to break out. His plan was to ferry his men to the north shore of the York River, where the American forces were the most vulnerable. To get his best soldiers into position, he planned to use sixteen flatboats, each making three round trips. Each trip would take about two hours. At 11:00 P.M., the first boat set off, crossed the river, and returned. "But at this critical moment," Cornwallis would later write, "the weather went from being moderate and calm, changed to a most violent storm of wind and rain and drove all the boats, some of which had troops aboard, down the river."

With some of his best troops on one side of the river, some on the other, and some floating away, Cornwallis lost all hope of breaking out. The next day the British raised the white flag. It was a heavyhearted General Cornwallis who wrote to General Henry Clinton, en route to relieve him, "I have the mortification to inform your Excellency that I have been forced to give up the posts of York and Gloucester and to surrender the troops under my command, by capitulation on the 19th inst. as prisoners of war to the combined forces of America and France."

George Washington also wrote a letter—to the president of the United States. Contrary to popular belief, George Washington was not the first president of the United States. The man George Washington addressed as the president was John Hanson of Maryland. After the Articles of Confederation were drafted, John Hanson had been officially elected "President of the United States in Congress Assembled." General Washington sent the letter announcing the victory to "John Hanson, President of the United States."

At the surrender ceremony, Cornwallis asked the band to play "The World Turned Upside Down." An American officer wrote, "The British officers in general behaved like boys who had been whipped at school. Some bit their lips, some pouted, others cried. Their round, broad-brimmed hats were well adapted to the occasion, hiding those faces they were ashamed to show."

The Revolutionary War not only created the United States; it also changed the landscape of another North American nation. After the war, about fifty thousand British loyalists emigrated to Canada. The British officially recognized American independence in 1783, but the first U.S. presidential election did not take place until February 4, 1789. On April 30, 1792, General George Washington took the oath of office on the portico outside the Senate chamber of Federal Hall in New York City and became the first president of these United States. What the weather was like that day is anyone's guess—no one wrote that information down at the time. The first recorded mention of the weather came sixty-five years later by someone who had not even been born at the time. According to Rufus Griswold, who got his information from Washington Irving, the sun was shining brightly as Washington took the oath. The only other witness to write anything about the weather was Mary Hunt Palmer, the daughter of a prominent general, who was a little girl at the time. When she was eighty-three years old, in 1858, she recalled that it was pouring rain at the first inauguration and that Washington carried an umbrella.

Hail to the French Revolution

If you want to keep control of a nation, it helps if the people are well fed. Perhaps no historic event dramatizes this principle better than the French Revolution. The storming of the Bastille, the drafting of the Declaration of the Rights of Men, the overthrow of Louis XVI and Marie Antoinette and their eventual executions—all would have far-reaching effects throughout Europe, and none of it might have happened were it not for a drought and a violent hailstorm.

The spring drought of 1788 could not have come at a worse time for the kingdom of France. The country was already suffering an economic crisis because of the debt it incurred helping the American colonists in their war against England, nor had the nation ever fully recovered from another major drought three years before. That time around, the peasants were left without enough grain to feed their livestock, and they had to slaughter many of their cows and horses, which led to a manure crisis. Waste from livestock was an important agricultural product used to fertilize the fields, and without it farmers were unable to eplenish the nutrients in the soil. As a result, many fields lay fallow.

At the time of the revolution, France had the largest population in Europe. Nearly 90 percent of the population was made up of the Third Estate, that is, the people who came after the first two estates—the nobility and the clergy. They farmed small fields; most of them had less than two and a half acres. Their diet consisted almost entirely of bread, which is hard to make when the fields are not producing grain.

In 1788 crops were failing across the countryside, and the cost of bread started to skyrocket. Tenant farmers who had been spending a little more than half their income on bread were now paying as much as 85 percent of it. Meanwhile, they were still being asked to pay feudal dues to noble landlords, tithes to the Church, and taxes to support the king and queen in a lavish lifestyle. As if that weren't bad enough, the Church and nobility were exempted from most taxes. This led the author Voltaire to quip, "In general, the art of Government consists in taking as much money as possible from one part of the citizens to give it to another."

Then the final insult to the crops came hurtling from the sky. On July 13, 1788, a hailstorm with stones sixteen inches (40 cm) in circumference pounded the farms. The tempest was described by Lord Dorset, British Ambassador Extraordinary to France, in a series of letters to his foreign secretary. His report, says J. Dettwiller, author of *Historique: La Révolution de 1789 et la météorologie*, may have been the longest meteorological report in diplomatic history.

> The noise which was heard in the air previous to the falling of the immense hail-stones is said to have been beyond all description dreadful. . . . The Hail-Stones that fell were of a size and weight never heard of before in this Country. . . . Not far from St.

> Germains two Men were found dead upon the road, and a horse so much bruised that it was determined to kill him from a motive of humanity to put an end to his misery; it is impossible to give description in detail of the damage that has been done. . . . A Country of at least thirty Leagues in circumference is entirely laid waste, and it is confidently said that from four to five hundred Villages are reduced to such great distress the inhabitants must unavoidably perish without the immediate assistance of Government; the unfortunate sufferers not only lose the crops of the present year but of three or four years to come, the vines being entirely cut up. . . .

The only thing that could make the situation worse would be if the hungry people of France were forced to endure an especially cold winter. This is exactly what happened.

In 1788, as Thomas Jefferson recorded in his autobiography, "there came on a winter of such severe cold, as without example in the memory of man, or in the written records of history. The Mercury was at times 50 below the freezing point of Fahrenheit and 22 below that of Réaumur. [The Réaumur temperature scale was devised in 1731 by French scientist René Antoine Ferchault de Réaumur. On the scale 0 represented freezing and 80 was the boiling point. –22° Réaumur is equivalent to –45°F] All outdoor labor was suspended, and the poor, without the wages of labor were, of course, without either bread or fuel."

The cold winter caused such heavy ice on the rivers that water mills could not operate, and grain could not be transported. The French government was forced to take emergency action and erect hand mills in Paris to grind flour. It was so cold that wine froze in the caves and barrels shattered, causing further problems for the vintners who had lost crops in the hailstorm and who were experiencing slow wine sales because of the economic recession.

As bread became more and more scarce, the hungry populace became increasingly angry and prone to acts of violence. Throughout the winter of 1788 and into spring 1789, they lashed out against the fixing of grain prices, forcibly opened granaries, and destroyed documents relating to the feudal obligations of peasants. Some bakeries were forced to close out of fear of vandalism, looting, theft, or worse. Soldiers were sometimes called in to protect bakers from bodily harm, but the greatest outbursts were yet to come.

The bourgeoisie sensed, in the unrest among the peasants, a force that could be used for their political aims. It was the middle-class lawyers, doctors, artists, and bankers who sought to abolish some of the privileges granted to the nobility and the Church. They circulated pamphlets, called *libelles*, to spread vicious rumors about the royal couple—Marie Antoinette was widely caricatured for her excesses. The English term "libel" derives from these revolutionary tabloids.

"This spirit of reading political tracts, they say, spreads into the provinces so that all the presses of France are equally employed," wrote Arthur Young in June 1789. "Nineteen-twentieths of these productions are in favour of liberty, and commonly violent against the clergy and nobility . . . inquiring for such as had appeared on the other side of the question, to my astonishment, I find there are but two or three that have merit enough to be known."

Most people, if they know nothing else about Marie Antoinette know her infamous response to the bread shortage: "Let them eat cake." The thing is, she never actually said it. In his *Confessions*, the French political writer Jean-Jacques Rousseau recorded the incident and attributed it to "a great princess." The book was written in 1767, three years before Marie Antoinette came to France, though it was not published until the final years of Marie Antoinette's reign. Yet because it fit in nicely with the picture of the queen as painted in the *libelles*, the statement became associated with her.

In May 1789, the States General, made up of the clergy, the nobility, and the Third Estate, met in order to find a solution to the problems of national debt and civil unrest, but the meeting was not fruitful. The electors were divided over whether to give each person a vote, which would favor the commoners, or to vote by estates, which would allow the clergy and nobility to outvote the commoners. The Third Estate grew tired of this bickering and decided to form a new governing body in defiance of the king. The organization met in the jeu de paume court and vowed not to leave until France had a new constitution. The king had no choice but to reluctantly allow the three estates to form a representative legislative body, the National Assembly; but his intention was to dissolve it as soon as he'd built up enough troops.

By July, mistrust of the aristocracy and anger over the bread shortage had reached a fever pitch, and it had taken on definite political overtones. It was called the Great Fear, because rumors of an aristocratic conspiracy had driven the hungry peasants nearly to a panic.

"The worst time after a bad harvest was always the early summer," wrote A. Cobban in the 1957 *A History of Modern France*. "The produce of the previous year's harvest was exhausted and the new harvest had not yet been brought in."

A year after the devastating hailstorm, bread reached its highest price ever. It peaked on July 14, 1789, the day the Bastille was stormed. The French would find a new way to water their crops, in the bloody words of "La Marseillaise": "Marchons, marchons / Qu'un sang impur / Abreuve nos sillons." (March! March! Let impure blood water our furrows!)

With the fall of the Bastille, that hated symbol of royal tyranny, the unrest became a full-fledged revolution. It was the beginning of the end of the reign of Louis XVI, and of the mystique of European kings.

Rain Ruins Robespierre

January 21, 1793, was gray and foggy in Paris. In the center of a public square, named for Louis XV, was a huge empty pedestal. Only a year before, that pedestal had supported a statue of the king, for which the square was named, seated heroically upon a horse. But the people of France had no more reverence for kings; they wanted no heroic images of them, and no reminders of the old order.

Now, where that statue once stood was a new edifice, one that the assembled crowd of twenty thousand could not know would become symbolic of the next era of French history: the guillotine. The people were there to witness the execution of a ruler whom they had once called the "restorer of French liberty," Louis XVI. At 10:22 A.M., when the guillotine blade fell, the death knell was sounded throughout Europe for the monarchy as it had heretofore been known.

With revolutionary furor sweeping France, being dead was not even enough to spare a royal. The magnificent royal tombs of Saint-Denis were smashed to bits; the coffins were opened, and the bones sold as souvenirs. In their zeal to remove all vestiges of the past, the revolutionaries would change the calendar that had ruled their lives, renaming the months of the year for the harvests and seasons. Their other new system of measurement, the metric system, would last a bit longer. One member of the Paris Commune called for the killing of all rare animals in the museum of natural history; another wanted to burn the national library.

The end of the old order, however, did not mean there was consensus as to what the new order should be, nor did the execution of the king put an end to the scarcity of food and rising prices. The country needed a strong personality to create a sense of order. Out of this power vacuum emerged Maximilien Robespierre. He would go down in history as one of the world's bloodiest dictators, in the words of author Eugene Methvin, "the first of many in history who mass-murdered in the name of liberal idealism." But what nature giveth, nature taketh away. An ill-timed (from Robespierre's perspective) downpour would prevent the man most associated with the Reign of Terror from escaping his own date with the guillotine.

As a young law student, Robespierre had been greatly inspired by the writings of Jean-Jacques Rousseau, who argued that people were inherently good but that they had been corrupted by civilization. Robespierre fully believed that once they threw off the yoke of the monarchy, the people would create a utopian state where liberty, equality, and brotherhood would reign. His passion for social issues and his ability to stir crowds with his oratory won him the nickname the "Incorruptible."

Robespierre was elected to the States General and became an influential member of the Jacobins, the political group that advocated universal suffrage, popular education, and the separation of Church and State. In 1792 he was elected to the Commune of Paris. As a deputy from Paris in the National Convention, he demanded the execution of the king and was instrumental in finally purging the assembly of the conservative Girondists, who had opposed the execution.

On July 27, 1793, Robespierre took his place on the Committee of Public Safety, which had been set up that previous April. The committee was faced with uniting a deeply divided nation, which threatened to fall into civil war at any moment. Robespierre wrote in his diary that what the country needed was "one single will." The National Convention would shape that single will, by any means necessary.

"Under a constitutional regime, little is needed but to protect the individual citizen against abuse of power by government," Robespierre said. "But under a revolutionary regime the government has to defend itself against all the factions which attack it. . . . Only good citizens deserve public protection, and the punishment of the people's enemies is death."

Those whom the committee considered enemies of the people grew and grew. There was, for example, the clockmaker, executed for "being arrogant, and speaking ironically about the revolution." In order to streamline the process of quieting these dangerous upstarts, the convention did away with the right to defense by those accused of a crime in the Law of the 22 Prairial. (Prairial, in case you don't recognize the word, is a month—the ninth month on the Revolutionary calendar.) Now accusing someone of a crime was as good as finding him guilty, which greatly improved the efficiency of sentencing people to death. Among the first executed under the new law were those who had argued against it.

On April 5, 1794, Robespierre used the power granted to him by the new law to silence his greatest rival, Georges Danton. Once a dear friend of Robespierre's, Danton is often credited as the chief force in the overthrow of the monarchy and the establishment of the First French Republic. In March 1794, when Danton's wife died, Robespierre wrote to him, "I love you more than ever, and I shall love you till I die." But after the initial revolution Danton failed to fall in line with the single will of Robespierre's people. He called for moderation and an end to what was coming to be called the Terror. Robespierre fell out of love, and Danton lost his head. Over the next three months, 2,085 enemies of the people would lose their heads to the guillotine. A special gutter was dug beneath the scaffold to carry away the river of blood from the daily executions. No one could be sure that he would be spared an accusation that would end his life.

By Thermidor 1794 (that's July to you and me), an underground opposition had developed within the Committee of General Security, which governed the police, courts, and prisons. Had Robespierre acted with the same resolve he had used against other enemies of the people, he might have saved his rule, but he became paralyzed with indecision. As his supporters urged him to strike down the conspirators, he holed up in his home, sulking, seemingly depressed by the betrayal.

He finally emerged on 8 Thermidor (July 26) and addressed the Convention. "We assert that there exists a conspiracy against public liberty. That it owes its strength to a criminal coalition which intrigues in the very bosom of the Convention. That this coalition even has accomplices in the Committee of Public Safety, and that some of its members are engaged in this plot. I know well who these calumniators are."

The conspirators had been signaled. Accuse or be accused—now was the moment. They had to act before Robespierre called for their executions. With the conspirators working overtime, Robespierre's supporters at the Jacobin Club urged him to use force and seize control of the Convention, but Robespierre believed the Convention was still on his side.

The next day, 9 Thermidor (July 27) was stiflingly hot and humid, the type of weather that makes tempers short. The Convention hall was crowded with Robespierre's Jacobin cheering supporters. One of Robespierre's faithful, Saint-Just, took the rostrum, but he was pushed aside by Jean-Lambert Tallien, one of the conspirators. His interest was personal—his mistress, Therezia Cabarrus, had been arrested along with her friend Josephine de Beauharnais, the future wife of Napoleon.

The emboldened conspirators called out, "Down with the Tyrant." Each time the Incorruptible tried to speak he was shouted down. Then a motion was made for Robespierre's arrest—at first, only a few supported it, but the cries gained momentum and an order for the arrest of Maximilien Robespierre and "his gang of scoundrels" was read, and he was taken away to prison. Then the Convention made the surprising decision to take a dinner break.

The Jacobin Club members were not about to sit and watch Robespierre go down without a fight. Heavily armed members of the Paris Commune rushed to free their leader. Paris seemed to be, once again, on the verge of civil war. It was the moment for Robespierre to unite his beloved people with fiery rhetoric, but he was dispirited by the rebellion. He had to be urged out of his cell by his rescuers.

Meanwhile, the members of the Commune gathered an impressive artillery and three thousand militiamen gathered, awaiting instruction from Robespierre, who arrived at the Hôtel de Ville at about 10:30 P.M. Rather than addressing the crowd, he and his closest advisers ignored them, concentrating instead on drafting a proclamation calling for assistance from the Commune. The members of that body, standing outside Robespierre's window, grew more and more unruly as they tired of waiting for a message from the Incorruptible. The "single will" of the crowd started to unravel. Fights broke out, windows were smashed, shops were looted, and some of the supporters ran away.

Robespierre could probably have regained control of the situation by simply addressing his supporters from the open window. Instead, nature intervened. At about midnight it began to rain—a few drops at first, but the shower soon came down "in trombs and torrents." The deluge continued for more than an hour. Lashed by rain and standing ankle-deep in puddles, even the heartiest supporters eventually gave up and went home. When Robespierre finally pulled himself together and looked out to address the multitudes, there were none: the square was empty. His last chance to retain power had been washed away.

Around 2:00 A.M. the Conventionists set out to recapture their infamous prisoner. With no crowd to stop them, they marched straight into the Hôtel de Ville. There they found Robespierre, surrounded by perhaps a dozen followers. He had just started to write his signature on the proclamation. Someone shouted "Vive Robespierre!" and a scuffle broke out. Robespierre could not bear to see his dream for the nation end this way. He turned his pistol on himself, but somehow he missed and only managed to shatter his jaw.

The next day, weak from the loss of blood, Robespierre was carried to the guillotine on a stretcher. Barely conscious, he was strapped down. The executioner ripped the bandage from Robespierre's shattered jaw. His last words were a scream of pain.

The United Irishmen, the French, and the Rain

By the late 1700s, the cries of "Revolution!" were inspiring writers, freethinkers, and rebels throughout Europe, and the kings and emperors were getting very nervous. The ideals of the American and French revolutions inspired a group of Irishmen to form the Society of United Irishmen in 1791. Their aim was to unite the Catholics and Protestants to throw off British rule.

At first, their plotting involved nothing but talk—shaping opinion, writing books. But when France declared war on Britain in 1793, one of the leaders of the United Irishmen, Wolfe Tone, saw an opportunity. Irish fighters would unite with the French and plan a full-scale invasion, driving the British off Irish soil. On December 15, 1796, this almost came to pass, but a storm intervened and dispersed the French fleet.

Theobald Wolfe Tone was born in 1763, the son of a coach maker. He graduated from Trinity College, Dublin, with a law degree and joined the Irish bar in 1789. Although he was not Catholic himself, and had, in fact, never met a Catholic, he published an influential "Argument" on behalf of the Catholics of Ireland and became one of the best-known advocates for the Catholic cause. Yet Tone, as it happens, disliked priests, the papacy, and religion in general. He believed that all religions would eventually be overthrown after the people won their political freedom; however, he also believed that discrimination against Catholics caused anger that kept the Irish divided, and that this kept Ireland from obtaining its independence.

On the strength of his "Argument" Tone was appointed assistant secretary to the Catholic Committee. While working for the committee, he came into contact with Archibald Hamilton Rowan, who had proposed the creation of an Irish equivalent of the National Guard of Republican France. That plan was almost immediately suppressed by Dublin Castle. Through Rowan, Tone met William Jackson, a former Church of England curate who was actually a French spy sent to Ireland to find out how the Irish were likely to react to a French invasion of their country. The idea intrigued Wolfe Tone, who gave his response in writing. "The misery of the great bulk of the nation, and above all, the hatred of the English name resulting from the tyranny of nearly seven centuries, there is little doubt that an invasion in sufficient force would be supported."

Unfortunately for Tone, a copy of this opinion made its way to the English authorities. Jackson was arrested, brought to trial, and found guilty. Before he could be sentenced, he poisoned himself with arsenic and died in the courtroom. Rowan escaped to France, and Tone was encouraged—as many Irish rebels of the time were—to seek his fortune in the United States.

It was a bitter, angry Tone who sailed to America with his family in June 1795. He was more determined than ever to put Jackson's plan into action. As soon as he arrived in Philadelphia, he sought out the French minister, Pierre Adet, and asked for France's help to liberate Ireland from the English. Tone left the United States, bound for Paris, on January 1, 1796. He arrived at the beginning of February with a coded letter from Adet addressed to the French Committee of Public Security. The charismatic Tone set about lobbying the French authorities for military aid. Although he was well received, he was not having any luck in his quest until he met Lazare Hoche.

Hoche was the commander of the Armée des Côtes de l'Océan, which had been involved in an insurrection along the western seaboard in Brittany and the Vendée. The campaign had cost the lives of six hundred thousand, and Hoche was anxious to face the English again. Hoche's plan involved a fleet of fifteen thousand and Tone himself would sail on board the *Indomptable* with General Emanuel de Grouchy as a chef-de-brigade.

On December 16, 1796, a fleet of seventeen ships set sail from Brest. The original plan had called for more ships, but the others were unable to join the squadron because of the weather. The commander of the fleet, Admiral Justin Morard de Galles, was afraid of running into the British at sea, and he adopted evasive tactics. Thanks to a dense fog, eight of the ships got separated from the rest of the fleet. It was only luck that brought them together on December 21 off Bantry Bay. The plan was to anchor off Bere Island, but what Tone described as "an infernal easterly wind" kept most of the ships out at sea. The storms continued for nearly two weeks, kicking up a greater and greater fury. "It blew a perfect hurricane," Tone wrote. General de Grouchy couldn't get any of his men ashore. What is more, the United Irishmen, who were supposed to fight alongside the French, stayed in their dry homes. Frustrated, tired, and very wet, the French fleet called it quits and headed back home.

We will, of course, never know what would have happened if the skies had been clear and the French had been able to come ashore. The French had promised they would help the Irish as they had assisted the Americans in 1778. They would take part in the battles, then leave the ruling to the Irish. Not everyone, however, was convinced that this was what they had planned.

"The strategy of the French alliance was itself extremely dangerous," wrote historian David Wilson in *The United Irishmen, United States: Immigrant Radicals in the Early Republic.* "Armies of liberation had a strange habit of becoming armies of occupation, and the 'sister republics' that France established on the continent looked suspiciously like satellite states. . . . Dependence on the French threatened to replace one form of imperialism with another."

Instead, the close call fueled English fears and republican rumblings. The United Irishmen intensified their revolutionary plans, and the Protestant loyalists became increasingly determined to nip them in the bud. The backlash was fierce, and it fueled Catholic anger, which created more militants.

In 1797 Tone tried again to stir French interest in an Irish invasion, but the plot never regained its momentum. Napoleon Bonaparte was not particularly interested. The United Irishmen made plans for a new rebellion in 1798 and expected French participation on the scale of the aborted attack, but Tone was able to stir up only enough forces to make small raids on various parts of the Irish coast. Only in County Wexford did the rebels make any gains, but the rebels were soon defeated at Vinegar Hill on June 21, 1798. When the French finally arrived in force, it was too little too late—the rebellion was almost over. The French force won a victory at Castlebar, but it was soon surrounded and captured.

In September Tone was captured in Donegal. At his trial he declared that his goal was "in fair and open war to produce the separation of the two countries." He was sentenced to be hanged, but before the sentence was carried out, he cut his own throat with a penknife and died on November 19, 1798.

The United Irishmen's rebellion would make an impact not only on Ireland but also on the cultures of the United States and Australia. Many of the rebels fled to America or were sent off to the penal colonies of Australia.

A Slave Revolt Washed Away

August 30, 1800, might have been remembered as the day that thousands of slaves in Richmond, Virginia, rose up against their masters, took the city armory, killed any whites who resisted them, marched into nearby towns, freed all the slaves, and made Virginia a homeland for displaced Africans. It might have been, had it not been for a storm that flooded bridges and roads.

That slavery was more prevalent in the American South than in the North was not simply a matter of ideology. The nation was initially divided into slave and free states largely because of the weather. The problem with owning people is that you have to house and feed them, whether they are working or not. This is not such a conflict in an area with a long growing season. But in the chilly North, the summers were not long enough for the profits from slave labor to outweigh the high cost of maintaining indentured servants.

Thus most slaves ended up in Brazil, the Caribbean islands, and the American South, where the tropical temperatures and long growing seasons made slave ownership profitable. By the Revolutionary War period, 40 percent of African Americans in the northern states were free, compared to only 4 percent in the southern states. A little-known fact is that African slavery was such a vibrant trade, the forced migration of Africans exceeded that of Europeans to the New World until the 1830s. The cumulative total of African migrants continued to exceed that of Europeans until the 1880s.

Eventually, people noticed the North/South slave divide, and tried to come up with a justification or explanation. Many correctly surmised it had something to do with the weather. Their conclusions, however, were often well off the mark. Many in both North and South came to believe that something about the southern climate was hazardous to whites, but Africans were immune.

Governor Johnson of Georgia summed up this point of view in an 1850 speech: "They cannot hire labor to cultivate rice swamps, ditch their low ground, or drain their morasses. And why? Because the climate is deadly to the white man. He could not go there and live a week; and therefore the vast territory would be a barren waste unless Capital owned labor."

The weather was also often blamed for slaves' "poor work habits." Slave owners dismissed out of hand the notion that people who have no share in the wealth, no prospects for their future, and who are treated as property might be less than enthusiastic about their work. Instead some thinkers of the time blamed the air. "When disposing themselves for sleep," wrote Dr. Samuel Cartwright of Louisiana, "the young and old, male and female, instinctively cover their heads and faces, as if to insure the inhalation of warm, impure air, loaded with carbonic acid and aqueous vapor. The natural effect of this practice is imperfect atmospherization of the blood—one of the heaviest chains that binds the Negro to slavery."

While the northern states did not support slave ownership, their economy was also dependant, albeit indirectly, on slave labor. Cotton, grown by slaves, was the nation's leading export in the early nineteenth century. Low-cost cotton was needed in New England textile mills, and exports to England of the raw material and the textiles powered much of the nation's economy as a whole.

In 1807 the United States passed legislation banning the slave trade, which would take effect in 1808. That same year the British Parliament outlawed the forced migration of Africans, and in 1810, the British negotiated an agreement with Portugal calling for gradual abolition of slave trade in the South Atlantic. But while the African slave trade ended, the institution of slavery did not. Thus the value of existing slaves began to skyrocket.

This meant that even in temperate southern states, it was no longer economically feasible to support a slave year-round. To make up the difference, plantation owners would rent out slave labor during slow seasons, and increasingly young slaves were taught trades to increase their value. Unlike the previous generation of slaves, these skilled tradesmen learned about life off the farm. They interacted with city people and slaves from other plantations. This would prove to be quite dangerous to the institution of slavery.

One such slave-tradesman was a tall, charismatic twenty-four-year-old named Gabriel. He is sometimes recorded as Gabriel Prosser, but "Prosser" was the last name of his owner, not of his parents. He was one of three sons born into slavery on the plantation of Thomas Prosser and his wife, Ann. Someone, perhaps Ann, taught young Gabriel to read and write, making him part of the very small minority, about 5 percent, of slaves with this ability. In order to make the best use of his time, they provided him with training as a blacksmith.

"In a different world, Gabriel would have prospered," wrote historian Douglas R. Egerton. "His intelligence, his physical size, and his skill would have marked him as a man on the rise. But this was not a different world; Gabriel was a black man in Jeffersonian Virginia."

He could have done what many in his situation did—run away to a free state. But Gabriel did not want to abandon his brothers or his new wife, Nanny. In any case, running away was abandoning responsibility. Gabriel did not just want his own freedom; he wanted freedom for all slaves. He began to believe that the only solution was to rebel against his oppressors. As he listened to political discussions and read newspaper headlines, he became more convinced that the time to act was approaching. The nation was deeply divided; the Union itself seemed ready to split.

Around this time Gabriel met a Frenchman named Charles Quersey. The abolitionist was notorious among plantation owners. He traveled through the slave states suggesting that blacks rise up against the whites. Gabriel became all the more convinced that now was the time. His plan was that Charles Quersey would help him organize the insurgents, and together they would amass an army of one thousand men and march into Richmond. Such an army would have been quite possible, as about eight thousand blacks lived in the Richmond area. A small percentage was all they would need to overpower the whites if they were well organized. They would set fire to the warehouse district as a diversion, take the treasury, and divide the money among the rebels, then hoist up a flag reading Death or Liberty. Finally, they would capture the governor, James Monroe, and ransom him for their freedom. Quakers, Methodists, French people, and poor white people were not to be harmed because Gabriel believed they possessed almost as little power as the slaves and that they would join the cause.

As a blacksmith, Gabriel enjoyed greater freedom of movement than many slaves. A few days each month, Gabriel worked on various plantations around Richmond. There he met and spoke to many other slaves. Little by little, Gabriel revealed his plan to a select group of friends, and the number of conspirators grew and grew. In secret, they fashioned crude swords and bayonettes out of farm tools over the course of several months. Everything was set to go on August 30. That morning, however, Tom and Pharaoh, two slaves belonging to Mosby Sheppard, warned their master of the plot.

Then, around sunset, as a local recounted, "there came on the most terrible thunderstorm, accompanied with an enormous rain, that ever I witnessed in this state." Roads were washed out, creeks rose, bridges were flooded, and travel and communications were effectively stopped. Only a small portion of Gabriel's men were able to get to the meeting place. The handful of soaked, muddy slaves would not be able to lead their resistance, and the way the water was rising, they'd be lucky to make it back into town. As best they could, the rebels spread the word that they were to meet the following night.

Mosby Sheppard, meanwhile, realizing he was on a death list, was motivated enough to brave the weather. He was not able to alert the militia captain, William Austin, but he did get as far as the local tavern, where he shared what he knew with his peers. The plantation owners sprang into action the next day by arresting dozens of blacks. Gabriel got away on a ship, but he was spotted in Norfolk a month later and brought back to Richmond to be tried. At the trial Gabriel, or perhaps another of the captured slaves, made a memorable speech.

"I have nothing more to offer than what General Washington would have had to offer, had he been taken by the British and put on trial by them. I have adventured my life in endeavoring to obtain the liberty of my countrymen, and I am a willing sacrifice to their cause; I beg, as a favor, that I may be immediately led to execution. I know that you have predetermined to shed my blood, why then all this mockery of trial." They granted his wish, and hanged him on October 8. Two hundred two years later, the City Council of Richmond unanimously passed a resolution calling Gabriel Prosser an "American patriot and freedom fighter."

Gee, It's Cold in Russia, Part II

Napoleon Invades Russia

When Napoleon set his sights on Russia in 1812, he seemed to be an unbeatable god of war. He planned to march over the Nemen River into Russia with the greatest fighting force ever seen: six hundred thousand men of the Grande Armée. The force included soldiers from areas already conquered or subdued by France, which was almost every nation in Europe. The conquest of Russia seemed all but inevitable—Russia's forces numbered only one hundred and eighty thousand. The czar, Alexander, had never led troops in battle. Napoleon had studied the ill-fated campaign of Charles XII of Sweden, but he believed he was much wiser than Russia's previous invader. With sufficient planning, the French would be out of Russia by winter, and attrition by weather would not be a factor. Nothing could thwart his ambition of being the "master of all the capitals of Europe."

What Napoleon had not fully counted upon was that Russia is not simply a land of harsh winters. It is a land of climactic extremes, as General Philippe-Paul de Ségur, Napoleon's aide-de-camp, would soon come to realize. "For such is the Russian climate," he would later write. "The weather is always extreme, intemperate. It either parches or floods, burns or freezes the earth and its inhabitants, a treacherous climate whose heat weakened our bodies as if to soften them for the cold which was soon to attack them."

The weather did not cooperate with the French from the start. As the offensive began, the soldiers were doused by severe thunderstorms. Napoleon's men had set off to capture Moscow with large wagons to cart equipment, weapons, and fourteen days of provisions for each soldier. Many of these wagons were soon abandoned as the wheels were buried in mud.

While much is made of Russia's particular brand of cold, fewer people realize that it also has hot. As Napoleon's army marched toward Moscow through Lithuania, summer heat devastated the ranks. They had no tents, so they slept out in the open whether it was dry or rainy. Their boots disintegrated. Wells were scarce, and some dehydrated men were reduced to drinking horses' urine out of ruts in the road.

The Russians, meanwhile, refused to engage the Grande Armée in a decisive battle. They were apparently determined to answer the question: "What if they gave a war and nobody came?" The French marched; the Russians retreated. The French marched farther; the Russians backed off. There is always plenty of Russia to retreat into. After two months of this, Napoleon's main force had been reduced by one hundred thousand—most killed by heat and exhaustion, not musket balls. It was starting to look as though the French could lose the war without ever fighting a real battle.

Napoleon was determined to score a victory before winter rolled in. In September his men fought the Battle of Borodino, the only full-scale engagement of the campaign. The battle was fierce, bloody, and indecisive. Then the war returned to its familiar pattern. The Russians withdrew and the French pursued them.

A few weeks later Napoleon's men arrived in Moscow. Their prize failed to live up to their expectations. The soldiers were looking forward to finding a few Russian women to entertain them. "For that is the character of the French soldier," wrote Sergeant Adrien Bourgogne. "From the fight to lovemaking, and from lovemaking to battle." Napoleon expected a Russian delegation to meet him, to hail him as the victor, and to give him the keys to the city so he could set down the rules for his newly acquired subjects.

Instead they found an abandoned city. "There was nobody there to give the men that to which they had a right, so what could the French soldiers do?" wrote the Saxon Sublieutenant Leissnig. They started looting. First, they simply sought food, clothing, and entertainment. Then they loaded up their satchels with gifts for their wives back home—candlesticks, picture frames, jewelry, dresses, and furs.

The Russians who remained in Moscow took a page from Peter the Great's playbook. As the last units of the army marched away, they set fire to anything that might be useful to the French should they arrive there. According to Roger Parkinson, author of *The Fox of the North*, as the Russians marched away from Moscow, a military band began to play in order to raise the soldiers' morale. General Mikhail Miloradovich rode up to the commander in a fury. "What idiot told your band to play?" The band officer explained that under the regulations set down by Peter the Great, a garrison must play suitable music when leaving a fortress. "Where do the regulations of Peter the Great provide for the surrender of Moscow?" Miloradovich shouted. "Order that damned music to be stopped!"

Behind them, the dispirited soldiers left a city in ruins. Stoked by the wind, the fire blazed out of control, consuming blocks upon blocks of wooden buildings, sending tin roofs and church domes flying. For three chaotic days the fire consumed everything in its path. When it finally died down, two-thirds of Napoleon's prize was gone. When it became clear that Czar Alexander I had no intention of surrendering, Napoleon ordered a retreat on October 18.

The weather was relatively mild throughout October and Napoleon openly scoffed at those who had regaled him with horror stories of the Russian winter. But that was all about to change. Napoleon was to learn firsthand what his advisers had been talking about. On November 6 a pouring rain turned to snow, which blanketed the ground. The country quickly became, in the words of the nineteenth-century historian Charles Morris, "a desert; a series of frozen solitudes incapable of feeding an army, and holding no reward for them. . . . It was a chill, inhospitable country to which the demon of war had come."

Cold is an equal-opportunity killer. The Russians, too, were battered by it, but they at least were dressed appropriately. In those days, armies tended to take a vacation from fighting in the winter, so the French did not have winter uniforms. Their uniforms did not even cover their stomachs, which were protected only by vests. The helmets of the dragoons actually drew the heat away. The falling French soon turned into an army of cross-dressers. The silks, furs, skirts, and liturgical vestments looted from Moscow became their only protection from the chill.

"It was a continuous masquerade, which I found highly entertaining and ribbed them about as they passed," wrote a Colonel Pelet.

The one upside to the freeze was that it hardened the ground. Carts slid easily over the ice, and many of the coachmen removed the wheels and created makeshift sleigh runners for them. Unfortunately, a couple of days later, the snow melted, the ground turned into muck, and the runners were useless. Many of the wagons had to be abandoned along with their rations, weapons, and luggage. More wagons were abandoned when the horses that pulled them fell down dead of cold. The animals then became a meal for the starving troops. When horsemeat seemed too gamey, the soldiers poured gunpowder on it to mask the flavor. Stray dogs and cats were no longer safe.

When the French happened upon a standing village, they ran the risk of burning it down themselves. Russian homes were heated by stoves made of wood rendered with clay. They had to be heated up gradually, but tell that to a man who is dying of starvation and frostbite. They fired those stoves up as fast as they could, and next thing you know, the stove would catch on fire, and the house would go up in flames, taking several soldiers with it.

On November 25, with about fifty thousand troops left, Napoleon reached the Berezina River. The Russians had destroyed the bridge that he intended to use to cross the deep, south-flowing river. The weather was just about the worst it could be. Had it been only a little colder, the river might have frozen over and the men could have simply marched across. As it was, the air was just cold enough to give the water a deadly chill, and to fill it with floating masses of ice, no good for crossing and treacherous to anyone who ventured into the water.

Given the size of Napoleon's army, the only way to get to the other side would be to build two bridges as quickly as possible. A few brave soldiers were sent into the water to their all-but-certain deaths in order to put the underwater supports into place. They worked continuously through the night, and by the morning of November 26, the Grande Armée was able to cross. It took more than a day to get the remains of the army across. When they were safely on the other side, they burned the bridges so the Russians could not follow.

The cold continued to cut down the French army. On December 6, temperatures dropped to –38°C (–36°F). The skeletal soldiers were reduced to "savage beasts." They would kill or be killed in order to wrest a piece of horsemeat or a coat from the back of a dead man. As many as forty thousand men perished in just four days—their bodies littered the streets. Back in the Lithuanian capital, Vilnius, the dying men are said to have ransacked local medical schools in search of preserved human organs to eat.

For months, the locals tried to clean up the dead bodies. The hardened ground prevented them from digging graves, so they threw the corpses into a defensive trench the French had dug at the beginning of their campaign. In 2002 bulldozers excavating for a new housing development uncovered their mass grave, the final resting place of two thousand men.

Of the six hudred thousand that marched into Russia, only thirty thousand marched back out. Most of the dead lost their lives to the elements, not to battle. Also abandoned in Russia were one hundred and sixty thousand horses and all of the Grande Armée's eight hundred cannon.

Those cannons proved useful seventy years later when Tchaikovsky wrote the *1812 Overture*. The piece has become a July Fourth standard with American orchestras (and many Americans vaguely assume it has something to do with the War of 1812), but it was written to commemorate Napoleon's defeat in Russia. Incidentally, the English have produced a version performed by sheep and chickens, which led author Margaret Atwood to quip: "Generals screw up, their fiascoes get made into art, and then the art gets made into fiascoes. Such is the march of progress."

Napoleon's tortuous and resounding defeat proved that the emperor was vincible. On December 17, 1812, *The Times* of London summed it up this way: "Whether Buonaparte escapes or not, his army is undone, and his military fame degraded. How he may face his people may be well worth this thought: that he may never enter again upon his career of blood, is well worth ours."

The Russian campaign is viewed as the beginning of the end of Napoleon's reign, which was sealed at Waterloo in 1814. It should have served as a cautionary tale for anyone intent on invading Russia. It didn't.

Does That Star-Spangled Banner Yet Wave?

The American national anthem, "The Star-Spangled Banner," is a poem about a lawyer's attempt to spot the American flag after a night of battle. Set to the tune of an old English drinking song (with a vocal range of an octave and a fifth, it helps to be a bit liquored-up to hit that note in the line about the "land of the free"), the verse most U.S. citizens know is in the form of a question. It chronicles the author's frustration at being unable to spot the flag through the dense morning clouds. It is only in later verses, the ones only *Jeopardy* contestants have memorized, that we discover that yes, indeed, the flag is still there.

"The Star-Spangled Banner" does not commemorate the Revolutionary War but one of the most bizarre wars of American history: the War of 1812. The reasons for the war were murky. It began when the British, more concerned with what Napoleon was doing in Europe, had already made concessions to American demands. The major battle of the war was fought after a peace treaty had been signed, but the soldiers hadn't gotten the memo yet. Both the Canadian and American capitals were burned with little military justification, and Washington may have been spared from complete ruin by the outbreak of a tornado.

After the Revolutionary War, the "land of the free" was far from being a superpower. The fledgling nation came out of the war deep in debt, with industries and resources burned up by the struggle. "The Americans began the war without any preparation," wrote historian W. E. Scull in 1899. "They conducted it on credit, and at the end of fourteen years, three millions of people were five hundred millions of dollars or more in debt." The thirteen original states were "united" in name only. The state governments were more established than the national government, and a long dispute over how much authority Congress had over states' affairs would not be satisfied until the Civil War. No one considered America to be a major player on the world stage.

The War of 1812 came about mainly because of grievances over British maritime practices during the Napoleonic Wars. Both Britain and France engaged in blockades of the European coast, which interfered with U.S. shipping. President Thomas Jefferson's solution, in 1807, had been the highly unpopular Embargo Act, which outlawed trade with warring countries. The settlers of Vermont were initially not the least bit concerned, since they had no ocean ports, but when they discovered that the embargo forbade trade with their neighbor, Canada, a brisk smuggling operation between the two countries began. One creative method was to build a shack on the top of one of Vermont's many hills on the Canadian border. After stocking the house with goods, the smuggler would remove a supporting beam so that the house would "accidentally" tumble down the hill into Canada. The act was repealed in 1809, but the Napoleonic Wars continued to interfere with American shipping.

The French agreed not to interfere with U.S. trade to England, but only conditionally. In fact, both the French and the English were about equal in their interference with American ships, but U.S. anger toward its ally in the Revolutionary War was much less than its anger at its old enemy. By the time Congress declared war on June 18, 1812, and the date when the news of the declaration got across the ocean, England had already revoked the policy that caused the dispute.

No matter. The war was on. The first order of business was something that today sounds like the punchline to a joke. The U.S. invaded Canada. The goal was to expel the British from the colonies to the north to ensure security, and perhaps expand America in the process. Much of the Canadian territory was settled by British colonists who had fled U.S. territory during the Revolutionary War. When this new war broke out, many of these settlers were unsure which side to support and they did their best to simply stay out of it.

The campaigns against Canada's allied British–Native American force were nothing but reverses and fiascoes. There were numerous skirmishes over the Great Lakes, and Detroit was captured by the British-Canadians, then retaken by the U.S. In April 1813, the Americans captured York (now Toronto) and burned it. In retaliation, the British attacked Washington.

As four thousand well-trained British troops, fresh from fighting Napoleon, marched into the city on August 24, 1814, civilians and soldiers alike fell into a panic. The forces defending Washington were poorly equipped, trained, and prepared. Most did not even try to fight. The entire defense of the capital amounted to a single volley of musket fire and the full list of British casualties read: one killed in action, three wounded. The redcoats easily marched to the Capitol but assumed it could not possibly be so easy. Surely the Americans were waiting in ambush. The British shot rockets through the windows and stormed the entrance, but the building was absolutely empty.

Finding themselves in the seat of democracy, Admiral Cockburn, the leader of the British force, took the Speaker's chair in the House of Representatives and put the question to a vote. "Gentlemen, the question is, shall this harbor of Yankee democracy be burned? All in favor of burning it will say aye!" And by unanimous vote, the abandoned city was set ablaze.

Next they marched to the presidential residence—also abandoned. (First Lady Dolley Madison had famously rescued Gilbert Stuart's portrait of George Washington on her way out.) After rummaging for souvenirs like James Madison's love letters to his wife, the British torched it as well. One of the militiamen who fled the city was a lawyer named Francis Scott Key. He and his wife, Polly, joined a parade of families headed toward Maryland, where the president had fled. By the end of the night the Treasury had also gone up in flames.

One branch of the government did have a defender. Dr. William Thornton, the superintendent of the Patent Office, stood up to the invaders. He told them that if they burned the office they would be no better than the barbarians who destroyed the ancient library at Alexandria. With a shrug, the English left the Patent Office and headed to a deserted fort to destroy the stores of gunpowder. Here the British managed to do what the Americans hadn't: thirty redcoats were killed in their own explosion.

But more British soldiers would die by a twist of nature than by any gunpowder, American or English. With Washington still ablaze, a tornado smashed through the center of the city. Had the British just waited, nature might have done away with America's governmental buildings all by itself. George R. Gleig, a British military historian, described the events.

> Of the prodigious force of the wind, it is impossible for you to form any conception. Roofs of houses were torn off by it, and

> whisked into the air like sheets of paper; while the rain accompanying it, resembled the rushing of a mighty cataract, rather than the dropping of a shower. The darkness was as great as if the sun had long set, and the last remains of twilight had come on, occasionally relieved by vivid lighting streaming through it, which together, with the noise of the wind and the thunder, the crash of falling buildings, and the tearing of roofs as they were stript from the walls, produced the most appalling effect I have ever, and probably ever shall, witness. It lasted for two hours without intermission; during which time, many of the houses spared by us were blown down; and thirty of our men, besides some of the inhabitants, buried beneath their ruins. Our column was as completely dispersed as if it had received a total defeat; some of the men flying for shelter behind walls and buildings, and others falling flat on the ground, to prevent themselves being carried away by the tempest; nay such was the violence of the wind, that two pieces of cannon which stood upon the eminence, were fairly lifted from the ground, and borne several yards into the rear.

The storm quenched many of the fires the British had set, but it also laid waste to the one building the British had spared. The roof of Dr. Thornton's Patent Office was blown away. In this state of utter confusion, the British commander, Major General Robert Ross, ordered the retreat, but not before they had taken a prisoner, one Dr. William Beanes, who had been arrested for looting and jailing British troops.

The displaced attorney, Francis Scott Key, went to defend Beanes. He was visiting the British fleet in Chesapeake Bay when the attack on Fort McHenry in Baltimore began. It was not one of the war's major battles; due to the poor accuracy of the British weapons and the limited range of the American guns, little damage was done on either side. Yet to at least one observer, Francis

Scott Key, it seemed as though the fate of the nation hung in the balance.

With the capital abandoned and ruined, it seemed quite possible that the young nation, only three years older than Key himself, could be reabsorbed into the British Empire. The weather was still stormy, and Key stayed awake all night, looking at the fort through a spyglass, trying to spot the American flag through the clouds and the darkness.

"At 5:50 it was officially sunrise," wrote Walter Lord in *The Dawn's Early Light*, "but there was no sun today. The rain clouds hung low, and patches of mist swirled across the water, still keeping the night's secret intact. But it was growing brighter all the time, and soon an easterly breeze sprang up, clearing the air. Once again Key raised his glass—and this time he saw it. Standing out against the dull gray of the clouds and hills was Major Armistead's American flag."

When Key saw the star-spangled banner, he was inspired to write a little poem on the back of a letter he had in his pocket. A Baltimore newspaper printed it and suggested it be sung to the tune of "To Anacreon in Heaven." It was printed up in broadsides and found its way into various newspapers, and it was instantly popular. Key did not live to see his creation sung in baseball stadiums across fifty states. He died in 1843, and "The Star-Spangled Banner" became America's official national anthem almost a century later, on March 3, 1931. Many opposed it as the national song—it is too hard to sing, too militaristic, and set to the tune of a pub song. There is also a certain irony in an anti-British anthem set to an English tune. (A line in one of the verses you don't know celebrates that "their blood has wash'd out their foul footsteps' pollution.") At least, unlike the American patriotic ditty "My Country Tis of Thee," it is not set to the melody of England's national anthem.

American history records the War of 1812 as a second War of Independence. It was the battle that proved the United States was a force to be reckoned with and not simply an abandoned colony. American gloating over the victory is somewhat diminished by the fact that most British students have never even heard of the War of 1812. In 1814, as the war wrapped up, they were much more concerned with what was happening in Belgium in a place called Waterloo.

The greatest impact of the war was on Canadian history. The Treaty of Ghent, which ended the war, did not address any of America's original excuses for going to war. It did, however, clarify the border between the United States and Canadian territory. Mass American immigration to Canada stopped as England made it difficult for Americans to get land grants and encouraged British immigration. After the war, many settlers who were previously ambivalent about their nationality—thinking of themselves as neither American nor Canadian—chose sides. Anger at the American invaders gave many a strong sense of being Canadian. There has not been another war between the United States and Canada. The nations that were most seriously affected of all were those of the Native Americans, as we shall see in the next chapter.

Tecumseh Is Lost in the Fog

Tecumseh is one of those rare names, honored by Native Americans and European Americans alike. Born near present-day Springfield, Ohio, Tecumseh was raised by an older sister after his father was killed by whites in 1774 and his mother left with part of his tribe to Missouri. He grew to be a unifying figure who convinced the Native tribes to stop warring against one another and to focus on their common enemy: the white settlers who were intent on grabbing their land. By all accounts he was a charismatic, articulate speaker. The whites compared him to the young Henry Clay. The writings of W. E. Scull in 1899 are typical of how Tecumseh is remembered in American history: "Tecumseh, a mighty warrior of mixed Creek and Shawnee blood, was one who dreamt the dream of freeing his people. With eloquence and courage he urged them on, by skill he combined the tribes in a new alliance. . . ."

Tecumseh died in a minor battle in the War of 1812, on a day so foggy the combatants could not distinguish friend from foe. No one ever knew who was responsible for the fatal blow. His death would change the course of Native American history, as no leader would emerge to take his place. Warring between Indians and

European Americans would continue after Tecumseh, but the Americans would never again see the Native Americans as a great threat. Indeed, until 1815 the word "Americans" was usually used to refer to Native Americans. After the war, it referred to European Americans.

"It is almost absurd that such momentous historic consequences should come from such an insignificant battle," wrote historian Erik Durschmied, "but such is so often the course of history . . . an unexpected patch of fog decided the destiny of the warriors of the forests and prairies of North America."

Initially, when the European colonists settled in New England, they enjoyed relatively peaceful relations with the natives. There weren't that many colonists in the beginning, and they were more interested in buying and selling fur than in taking over great tracts of land. This all changed, of course, as more and more shiploads arrived from England and the Netherlands. For nearly two hundred years, the land-hungry settlers and many of the residents of the coveted land fought terrible, bloody battles. It was all in keeping with the great European tradition of kingdoms battling other kingdoms over territory.

When they came to this new land, the Europeans looked upon the tribal leaders as kings. For example, they dubbed Metacom, of the Wampanoags, "King Philip." Most Native American tribes, however, did not operate like European kingdoms. Native societies were much more democratic than the European nations of the time. In fact, the Iroquois League is said to have influenced the colonists in the formation of their government—the great seal of the United States, with an eagle clutching arrows, is remarkably similar to the symbols of the league. Since monarchy was the only form of government the Europeans knew, they projected it onto all the people they met. It also made it much easier when they wanted to buy land. They could purchase a piece of real estate from a "king" with the sense that he had the authority to sell on behalf of the tribe. This led to quite a bit of confusion—take the time when the Dutch bought Manhattan, for example. They paid the Canarsees $24 worth of trinkets in exchange for the property. The only problem was that the Canarsees were native to Brooklyn. The Weckquaesgeeks, who lived on Manhattan, were confused when the Dutch started moving in, acting as if they owned the place, and the two groups battled on and off for years.

Metacom (King Philip), like Tecumseh, tried to unite the New England tribes to mount a coordinated defense. Before a plan could be put in place, however, three Wampanoags were executed and fighting broke out spontaneously. The Indians raided settlements in Massachusetts, and the colonists retaliated with brutal assaults on Indian villages. The ensuing war, "King Philip's War" was one of the bloodiest in U.S. history. About three thousand Indians and six hundred settlers were killed. Although those numbers may not sound like much, in proportion to population, casualties were greater in this than in any other American war.

For the first two centuries, the settlers and the colonists battled. Military expenses for war with the Indians took up 80 percent of George Washington's federal budget. In the 1700s as European powers battled for control of the American continent, various native nations united with different European nations. The Seven Years War, also known as the French and Indian War, pitted native tribes against other native tribes.

In 1791 Tecumseh was living at Keth-tip-pe-can-nunk. The name of the trading post was a bit much for the white settlers, who referred to it as Tippecanoe. The site was a hub of trade between whites and the Potawatomi and Kickapoo tribes. Inevitably, some of the European Americans started to think that Indiana and Michigan had a few too many Indians in them, and they noticed that they seemed to gather around the trading post.

They razed Tippecanoe in order to scatter the natives. Tecumseh and his brother, Tenskatawa, known as "the Prophet," were incensed. The brothers were determined to rally support to overcome the white menace. Tenskatawa rallied his people at home, while Tecumseh traveled throughout the land. He spoke to the people of various tribes and convinced them that the only way to reclaim their land and their freedom was to form a mighty alliance.

By 1811, Tecumseh had stirred up the passions of tribes as geographically diverse as the Seminole in Florida and the Iroquois in Quebec. He had also stirred the emotions of the whites but in a much different way. A united Indian uprising was the United States' worst nightmare. One tribe at a time they could quell, but the combined force of all of the Indian nations would be too much for the military to cope with. The newly appointed governor of the Indiana territory, General William Henry Harrison, knew all about Tecumseh. He had served as an aide to Major General Anthony Wayne at the Battle of Fallen Timbers near what is now Toledo, Ohio, on August 20, 1794, and among the Shawnee warriors had been Tecumseh. Wayne's American Legion had been victorious that day, and Harrison had every reason to believe he would be victorious again as he prepared an attack on Tippecanoe.

Harrison assembled a militia of a thousand men with muskets and descended on Tippecanoe on November 6, 1811, to set up camp. Tecumseh tried to warn his brother not to engage the U.S. forces in open battle. But Tenskatawa believed in the power of rhetoric and inspiration—he told his men that a musket ball could never harm a brave. The next morning the Natives attacked and took the Americans by surprise: sixty-two of Harrison's troops were killed. The Shawnee won the day, but a number of the braves learned firsthand that the Prophet was incorrect about musket balls. They felt betrayed and threatened to kill Tenskatawa. When they left Tippecanoe to fight one another, Harrison's men burned the city and claimed victory. Harrison would be lauded as the hero of Tippecanoe in his presidential campaign. His slogan was "Tippecanoe and Tyler too!" There were Tippecanoe products including badges, handkerchiefs, even shaving cream.

The battle of Tippecanoe did nothing to lessen Tecumseh's loathing for the United States. When the War of 1812 broke out, he and his united force of braves joined the British effort. Although the War of 1812 was ostensibly fought over shipping rights, one of the underlying causes was the British policy of supporting the fight of the Native Americans of the Northwestern frontier against American settlement in the British territory of Canada. If the war was really about shipping, it should have been most popular in New England, where they were most economically hurt by maritime blockades. Instead, the war was so unpopular in those regions that New England militias often refused to cross the border to fight. That meant that President Madison had to shift his gaze from Montreal to the Indiana–Michigan territory where the settlers were more enthusiastic about the war.

In the Battle of the Thames (or the Battle of Moraviantown to the Canadians) the American General William Hull, governor of the Michigan territory, faced off against the British forces of Colonel Henry Proctor and Indian forces under Tecumseh and his deputy, Oshawahnah, of the Chippewa. Their regiment was made up of five hundred representatives of the Shawnee, Ottawa, Delaware, Wyandot, Sac Fox, Kickapoo, Winnebag, Potawatomi, and Creek. The combined British-Indian force was one thousand strong, but they were still outnumbered by the Americans three to one.

The field of battle, near present-day Thamesvile, Ontario, had the Thames River on the right with a marsh running parallel to the river for about two miles. The strip of ground between them was swampy forest. The narrow ground would take away some of the benefits of Harrison's numerical superiority. Tecumseh's men waited behind trees on the edge of the swamp ready to ambush the U.S. forces as they passed. The night of October 4, 1813, was typical for Canada in the fall. The temperatures dropped below freezing and frost covered the trees and the ground. The water of the swamp, however, stayed at a constant temperature. As the sun rose, the surface of the swamp heated up faster than the surrounding air. Warmer water vapor rose, creating a cloud on the ground, in other words, fog. While most of the battlefield was clear, the swampy areas that concealed the Indians was blanketed in gray mist. Tecumseh's men could barely see the ends of their muskets.

An invisible American unit of five hundred soldiers on horseback rode straight past Tecumseh's planned ambush, and right past the British lines. They were able to turn and attack the British from their left flank. The British troops were cut down and surrendered in less than ten minutes.

The Indians could hear the sounds of the battle, even if they could not see it. They raced toward the clamor and nearly bumped into the American cavalry. The riders were as surprised by the arrival of the Indians as the British had been by the Americans' arrival. They jumped down from the horses and engaged in hand-to-hand combat. Even so, a few bayonets found their way into the guts of friends. Arms would appear suddenly from the haze, with hatchets in hand, desperately whacking away at a murky but deadly foe.

Out of the fog some of the braves spotted Tecumseh, with blood dripping from a head wound, fighting and shouting encouragement to his warriors. No one saw who fired the bullet that blew his chest open. His body was carried from the field and buried in a secret place where the whites could not defile him. Thus ended the life of the great warrior Tecumseh, and his dream of creating a large-scale Indian confederacy. Had he succeeded, perhaps there would be a separate Indian state or nation today.

Instead, the battle ended with thirty-three Native Americans killed and four hundred and seventy-seven British taken prisoner. The American force lost only fifteen people. After the battle, the Indians had less confidence in their British allies and were more reluctant to come to their aid. The British lost the war, and in the resulting treaty agreed to give up their alliances with Indian nations. The Indians would continue to battle with the U.S. government for many years, but they would never again be seen as a serious threat. The European Americans, now known simply as "Americans," were free to expand westward from sea to shining sea.

The Water of Waterloo

> If it had not rained on the night of June 17, 1815, the future of Europe would have been changed. . . . A few drops of rain mastered Napoleon. Because Waterloo was the finale of Austerlitz, Providence needed only a cloud crossing the sky out of season to cause the collapse of a world.
>
> —Victor Hugo, *Les Misérables.*

If you visit the site of the Battle of Waterloo today you will find pictures of Napoleon, Napoleon souvenirs, mentions everywhere of Napoleon. You'd be forgiven for forgetting that the victory went to the Duke of Wellington, whose name is much less prominent. Waterloo was not the battle that made Wellington; it was the battle that unmade the emperor.

There were many reasons for this loss, one of which was the weather. Rain showers on the night of June 17, 1815, softened the ground. This meant the French had to delay the morning's fighting until about 11:30, which allowed time for the Prussians to join the fray.

Early in 1814, Napoleon had been forced to abdicate and he was banished into exile on the island of Elba, off the coast of Italy. Apparently viewing the past fifty years or so as a failed experiment, the French Senate reinstated the Bourbon dynasty. Louis

XVI's brother became king. Thing is, the French people were not any more enamored of the idea of having a king than they had been before the revolution. Those who remained loyal to Napoleon wrote to him and asked him to come back, which he did on March 20, 1815, and began what came to be known as his "one hundred days' rule."

While he was cheered by crowds in Paris, his return was not nearly as welcome in the rest of Europe. Sir James Mackintosh summed up the English view when he said: "Was it in the power of language to describe the evil? Wars which had raged for more than twenty years in Europe; which had spread blood and desolation from Cádiz to Moscow, and from Naples to Copenhagen; which had wasted the means of human enjoyment, and destroyed the instruments of social improvement . . . the work of our fortitude is undone; the blood of Europe is spilled in vain."

All of the powers of Europe united against this threat: Great Britain, Prussia, Austria, and Russia started planning for an invasion of France. Napoleon was well aware that it was coming and in less than a month, he had prepared a force of one hundred and twenty-five thousand men of his own. The allies, meanwhile, had assembled two armies in Belgium. The Prussians had one hundred and sixteen thousand men under the command of Field Marshal Blücher. The Duke of Wellington commanded a mixed force made up of British, Belgians, Dutch, and Germans, which numbered about ninety-three thousand. The Austrians and Russians were to join them in July, and a force of six hundred thousand would march into France.

Napoleon had no intention of waiting, and he led his army to Waterloo. Two corps of the Prussian army were located near the village of Ligny with Wellington's force stationed near Quatre Bas, which was 12 km to the north. Napoleon split his army to deal with each force separately in order to keep them from joining. Marshal Ney, commander of his left wing, attacked Quatre Bas, and Napoleon led the rest of his forces to Ligny. They got a later start, which allowed only six hours of daylight in which to wage war. As they began the attack, a thunderstorm burst, which made it impractical to use muskets. The French charged with their bayonets. Soon blood mixed with the rain and soaked into the Belgian soil. The Prussians lost sixteen thousand in the engagement and were forced to retreat. Although the French won the day, they also suffered terrible losses—eleven thousand members of the Imperial Guard lay dead.

With the Prussian army out of the way, Napoleon set his sights on the Dutch–English contingent. The rain continued through the night and into the morning of the seventeenth. "Large isolated masses of thundercloud, of the deepest, almost inky black, their lower edges hard and strongly defined, lagged down, as if momentarily about to burst, hung suspended over us, involving our position and everything on it in deep and gloomy obscurity," wrote the British officer Lieutenant Hope in a letter.

Napoleon decided to postpone the beginning of his attack on Wellington until noon, when he hoped the ground would be dry. As he waited for favorable weather, Wellington heard about the Prussian defeat. If the battle had started before he received this intelligence, he might have been forced to engage the French full-on and wait for help that might never arrive. Instead, he decided to withdraw to a better position. When Napoleon got wind of Wellington's retreat, he had no choice but to follow. As he started the chase, the rain began to come down again, turning the field to mush. The soldiers had to crowd onto a paved road. This was slow going, and the French did not catch up with Wellington. Napoleon and his men made camp in the mud under torrential rains.

The skies finally cleared around 7:00 A.M. the next day. Napoleon hoped to get this battle going right away, but his commander, General Drouot, recommended he wait for a few hours to let the soil firm up. Not only would the muck hinder the cavalry, but the artillery shells would also be rendered ineffective by the mud. So once again the soldiers sat and waited. The killing finally commenced around lunchtime. The French were fairly routing their enemy, but around 4:00 P.M. the Prussians arrived and joined the battle. They, too, had been spared by the elements. The units that Napoleon had sent to find the Prussians and keep them from joining the other allies had been slowed by the sodden ground. The arrival of the Prussians changed the course of the battle, and Napoleon was defeated. An estimated twenty-five thousand French soldiers lost their lives and another eight thousand were taken prisoner. The Duke of Wellington lost fifteen thousand and the Prussian army lost seven thousand that day. Four days later, Napoleon abdicated for the last time. The expression "met his Waterloo" came to be applied to any crushing defeat in both the French and English languages.

Waterloo marked the end of France as a world superpower. The most populous country in the western world in the eighteenth century, it dominated international culture—its export of the ideas of democracy and revolution certainly took hold in the United States. French was the international language of diplomacy. Slowly, France was eclipsed by the nation it had had such a hand in creating, and its language, English, began to dominate. After centuries of fighting, the Battle of Waterloo also marked the last time the French and the English engaged in open war with one another. In every large war to follow, the British and French were allies.

Gee, It's Cold in Russia, Part III
A Senseless War Extended by the Weather

At the end of the Crimean War, an estimated one million soldiers and civilians lay dead. To put this staggering figure into perspective it helps to recall that in 1856, the earth's entire population was about one-sixth of what it is today. Not until World War I would more people die in a war. Very few wars evoke the same futility and senseless loss of life as the Crimean War. The official players were the Russians, Turks, French, British, and Italians, but it was another force—the weather—that would cause most of the deaths.

The two-year war began when young Czar Nicholas I attempted to expand the ever-shifting borders of the Russian Empire. During Nicholas's rule, parts of Persia, the Caucasus, and the Far East became Russian. As the Ottoman Empire started to crumble, Nicholas began to eye Turkey. In 1853 Russia invaded Moldavia (a region that is now divided between Romania, Moldovan Republic, and Ukraine) and Wallachia (now in Romania), which were then under Turkish control. Britain and France were afraid that Russia would go on to gain a foothold in the Middle East and dominate the region.

When it became clear that these European forces would attack, Russia withdrew from Turkey, but the czar hinted that Russia might need to occupy Constantinople to protect Orthodox Christians in the Ottoman Empire. So the British and the French declared war on Russia in January 1854. "Declaring war was one thing," wrote historian Richard Cavendish. "Finding somewhere to fight it was quite another."

The French and British forces eventually settled on the Crimea, a peninsula that extends from what is now Ukraine into the Black Sea. Annexed by Russia in 1783, the Crimea was an important base for the Russian Imperial Navy. The initial invasion of Allied forces comprised twenty-seven thousand British, twenty-five thousand French, and eight thousand Turkish troops. Although the number of Russians in Russia certainly outnumbered the foreigners, the size and scope of the nation meant that these troops were widely dispersed. The bulk of Russia's army was in the Gulf of Finland area protecting the approaches to the capital, St. Petersburg. Transportation has always been a challenge in Russia and in those days there was only one trail line into St. Petersburg, which extended only as far south as Moscow. The distance from Moscow to the Crimea had to be traveled in slow-moving carts and wagons. Thus, the Russian contingent in the Crimea was a modest fifty to sixty thousand. The first major battle, on September 20, 1854, was a victory for the invading Allies.

The next major battle, the battle of Balaklava, was an especially senseless exchange, immortalized in Lord Alfred Tennyson's poem "The Charge of the Light Brigade," which begins "Half a league, half a league, half a league onward . . ." and tells the heroic tale of brave soldiers who rode into "the valley of Death."

On the Allied side, this battle was headed up by two men whose names are recalled in fashion. James Thomas Brudenell, the 7th Earl of Cardigan, wrote Richard Cavendish in

History Today, "was undoubtedly badly spoiled. Handsome, hot-tempered, and lavishly be-whiskered, his three great passions in life were soldiering, hunting, and women." He was wealthy and vain and spent a portion of his wealth on new outfits for his soldiers. The splashy uniforms had tight crimson trousers that earned the regiment the nickname "Cherry Bums" and a knitted woolen jacket that buttoned down the front. (Hence the cardigan sweater.) Cardigan frequently did battle with his own officers, even wounding one in a duel. In 1854, when he was appointed the commander of the light brigade, he was, as Saul David wrote in *The Life of Lord Cardigan*, "the most unpopular man in England."

At the Battle of Balaklava, thanks to a misunderstood order, 676 riders of the light brigade were sent to attack a Russian battery at the end of a valley bounded on three sides by enemy troops. Although Cardigan questioned the ambiguous order from Lord Raglan (the inspiration for the sleeves), when it was repeated he followed it and led his men on the clearly suicidal mission. He is said to have muttered: "Here goes the last of the Brudenells." By the time the charge was finished only 195 of Cardigan's men were still mounted.

Despite the painful loss, Cardigan was thereafter heralded as a hero, thanks in no small part to Tennyson's poem. Interestingly, an original draft of the famous poem shows that the author was a bit more critical of the choice to send six hundred cavalrymen to certain death than he would later appear. Although Tennyson asked for the original drafts to be destroyed, two copies survived. As Poet Laureate, he was probably convinced by his friend Queen Victoria to make his account a bit more patriotic.

The Russians had their own poet in the field. While Leo Tolstoy was on the front lines and during the war, he had his first of many religious awakenings. After his service he wrote *Sevastopol Stories*, based on his war experiences. Whereas Tennyson exalted military heroism, Tolstoy debunked romantic ideas of martial bravery in his account.

And here, with each side reeling from a terrible loss, is where the weather takes center stage. Bolstered by their Balaklava victory, Prince Menshikov, commander in chief of the Russian army, planned his first major offensive for November 5. The state of the roads in Russia given the advanced season slowed the process of obtaining reinforcements: fresh troops did not get to the theater of war until November 4. Lord Raglan knew that reinforcements were massing but did not expect them to attack so soon. In fact, the Russians began their assault so quickly that the only good map of the area arrived from St. Petersburg *after* the battle.

The Russian forces consisted of two army corps commanded by Generals Pavlov and Soimonov under the command of General Dannenberg. They planned to attack an area known as Mt. Inkerman by the Allies and Cossack Mountain by the Russians. With only three thousand Allied troops, it was the most vulnerable. The Russians numbered about forty thousand.

Normally, of course, it is impossible to move forty thousand men, horses, and weapons without being noticed. On this day, however, a heavy fog set in. The Russians were able to tiptoe toward Mt. Inkerman with the British staring right at them, none the wiser. That night an inspecting officer looking over the valley said, "The night was more than usually quiet."

From the British perspective, there was nothing but a curtain of mist and rain that suddenly produced bursts of artillery before Russian soldiers appeared at their throats. Yet the fog that originally aided the Russians became their enemy. The men on both sides could see no more than a few feet around them, so there was no possibility to attack in orderly ranks; instead, it was a man-to-man brawl with bayonets in bellies. The fog kept the British from realizing how outnumbered they were or how many of their comrades had fallen. Thus their morale remained high, and they kept fighting until reinforcements arrived. The battle ended as a loss for the Russians, although it was truly a loss for all sides. The British suffered twenty-six hundred casualties; the French seventeen hundred, and the Russians, eleven thousand five hundred.

A few weeks later, the real war began—the full-on assault by the elements. The Russians were confident, sure that their secret weapon would provide the true reinforcement. They were not disappointed. On November 14, 1854, a storm began. Lord Raglan described it in a letter to the Duke of Newcastle as "the most violent tempest I ever witnessed, attended with thunder and lightning, heavy rain, hail, and later in the day snow, and the damage it has associated is fearful."

British tents were blown away, as were rations, blankets, chairs, and clothing. The ground was transformed into a marsh that would not allow soldiers to light their campfires. Hospital tents flew away, leaving the sick and wounded lying in pools of freezing water. Next came sleet and heavy snow that buried patients and soldiers. Several men died from exposure that night. The French commander got down in the freezing muck with his men to lift their morale. The class-conscious Lord Raglan and his staff stayed in a warm and comfortable farmhouse, generally unaware of the toll the storm was taking on the troops.

The British lost eleven ships, including the *Price*, which carried all the needed provisions, notably food, medical supplies, and winter clothing. A total of sixteen French ships, including the pride of the fleet, the *Henri IV,* were battered and sank. The loss of the *Henri IV* was a huge psychological blow to the French. Although Russian ships were also battered, most of them were anchored far from the center of the storm, so the damage was not as great. Interestingly, the event actually *raised up* a Russian ship. The ship, which had been deliberately sunk to block the entrance to Sevastopol's harbor, was dislodged by the storm.

When Nicholas I learned of the effect of the tempest, he wrote to Prince Menshikov, "Thank you for the storm, it has helped us greatly; another one would yet be desirable."

Without the supplies from their ships, the British and their allies were entirely unprepared for the winter. Although the Crimean winters were the mildest in Russia, they were far from being a tropical paradise and as fate would have it, the winter of 1854–1855 would be the coldest European winter on record. Vital supply routes were slowed for six weeks. In the meantime, the British covered themselves with pieces of uniforms of dead Russian soldiers and the hides of their dead horses. As the winter dragged on, they dug up the dead to take the blankets they had been buried in. Some men saw their shoes disintegrate, and their toes and feet soon succumbed to frostbite from walking barefoot in the snow. There are accounts of men thawing their beards over a fire so they could open their mouths. The ragtag uniforms made it hard to tell officers from enlisted men, and one regiment from another. The cavalry horses were also starving. With no greens to forage, the emaciated animals, too tired and sick to carry riders, wandered about trying to eat blankets and tent canvas. When the horses collapsed and died, the soldiers devoured them. The only animals that were growing fat were the vultures.

"None but the people of the hardiest constitution can stand it," wrote Colonel Sterling of the Highland Brigade. "All the others are dead or dying. I hear of men on their knees crying in pain."

As the men in the field froze or succumbed to diseases transmitted by the rotting corpses, Lord Raglan was rolling up his sleeves in the comfort of his warm farmhouse with a surplus of wood taken from the hulls of the smashed ships. Raglan managed to receive regular supplies of food, which were prepared for him by a French chef.

. . .

By January 21, the British could find only two hundred and ninety men healthy enough to man a trench line that was more than a mile long. By February, the Brigade of Guards, once the pride of the British army, had been reduced from three thousand to four hundred.

Although these weather events contributed greatly to the death toll from the Crimean War, they were not enough to result in a Russian victory. In January the Italian kingdom of Sardinia-Piedmont entered the war and sent ten thousand troops. In September the French had a major victory on the Malakhov, a major strongpoint in the Russian defenses. Finally, on September 11, 1855, the Russians evacuated Sevastopol.

Nicholas I died on March 2, 1855, and Alexander II became czar. He immediately began negotiating for an end to the dreadful conflict. Of course, these things take time, and it would be almost another year before the peace terms were finally agreeable to both sides. "We are not of those who regard the expedition to the Crimea as a mistake," said an editorial in the British journal *Economist*. "As it is, however, we must admit that it has turned out unfortunately. . . . We have lost 20,000 men and we have not gained enough land to make them 20,000 graves."

In the end 60 percent of the war's victims were done in by disease. Cold, clouds, and cholera would be the victors in a conflict that has been called "the world's most curious and unnecessary struggle."

"Perhaps the most fundamental lesson to be learned from a look back at the Crimean War," wrote historian Robert Edgerton, "is how easy it is for nations to blunder into wars that serve no purpose and cannot be won."

The war did have one positive consequence: it was the beginning of scientific weather forecasting. After the disastrous storm losses, Emperor Napoleon III asked the astronomer and mathematician Urbain Le Verrier to figure out a way to avoid such catastrophes in the future. After a great deal of investigation, Le Verrier concluded that a storm was a circulation of winds around a core of low pressure that moved over a steady and fairly predictable course. He established a network of weather stations connected in regular contact via telegraph with his Paris observatory and used the information to issue the first weather maps and storm warnings.

The Guy with the Sideburns Gets Stuck in the Mud

General Ambrose Everett Burnside was popular with his men, and something of a trendsetter. His side-whiskers, originally called "burnsides," give us the word "sideburns." He was also the namesake of the Burnside hat. When it comes to his Civil War battles, however, his name gets dragged through the mud. Burnside was the leader of an infamous episode remembered as the "mud march," a fiasco that was the end of Burnside's command.

At the end of 1862, morale was low among Union troops in the eastern United States. General George B. McClellan's attempt to take the Confederate capital, Richmond, Virginia, had failed in June. Two months later, the Confederate general Thomas "Stonewall" Jackson defeated the Union forces at the Second Battle of Bull Run. The battle of Antietam had ended in a draw with huge losses on the Union side. President Lincoln, under great pressure from Congress, fired McClellan and put Burnside in charge.

Burnside's first campaign was an unmitigated disaster. Burnside's strategy was to move the army to the south, cross the Rappahannock River, and take Fredericksburg, and from there straight to Richmond with an army of one hundred and twenty thousand troops. The Union general arrived on the northern side of the river in the third week of November, but he was unable to cross because his pontoon train had not yet arrived to put up a bridge. This meant that General Robert E. Lee was not the least bit surprised when the engineers finally got there and started building a bridge. Confederate soldiers took pot shots at the bridge builders while Lee assembled a force of seventy-eight thousand on the high ground behind Fredericksburg. Interestingly, they included two of General Burnside's cousins, Lieutenant A. W. Burnside, of the 3rd South Carolina Battalion, and Addison M. Burnside, of the 44th Georgia Infantry.

When the Union soldiers finally crossed, they were met with more gunfire from the Confederates, who were hidden in a sunken road behind a stone wall. The wall ringed a muddy field. Wave after wave of Union soldiers were shot down like ducks in a shooting range, but Burnside kept ordering men onto the field. The soldiers began to stack their own dead like cordwood for some kind of barricade. Accounts differ, but somewhere between eight thousand and twelve thousand Union soldiers were killed before Burnside ordered a reluctant retreat on December 14 under stormy skies. The mud had been a factor at Fredericksburg, but that was nothing compared to what Burnside would face on his second attempt to cross the river.

This time Burnside would head downstream in an effort to circumvent Lee's stronger force. The scheduled date for the attack was January 20, 1863. On the day in question the weather was perfect for the assault, but an intelligence report about Confederate positions south of the river resulted in a twenty-four-hour delay. Little did they know that a low-pressure center far to the southwest was ready to deal the Union a terrible blow. By noon on January 20, the sky was overcast. The temperature began to drop and the wind shifted and blew harder. Around 9:00 P.M. a raindrop fell—then another. The dotting of drops turned into a pelting torrent.

"If the storm had been smaller, or its rate of movement faster, or its path more to the north or south, the precipitation might have been light, fleeting, or avoided altogether," wrote Harold A. Winters, author of *Battling the Elements*. "But unfortunately for Burnside and his plan, this storm was very large, well developed, slow-moving, and now tracking directly over southeastern Virginia."

All night it rained, and by morning, the Virginia soil was what the locals called a loblolly—a thick gruel of mud. The soil in the area was, in the words of one officer, a reddish clay that held the water and became softer and softer.

"It appears as though the water, after passing through a first bed of clay, soaked into some kind of earth without any consistency," wrote a Colonel de Triobriand. "As soon as the hardened crust on the surface is softened, everything is buried in a sticky paste mixed with liquid mud, in which, with my own eyes, I have seen teams of mules buried."

Nevertheless, the seventy-five thousand demoralized soldiers started the march with a great mucky slosh. Their pace was so slow that they stopped after less than three miles and made a camp in the rain. They spent the next day mostly waiting for the artillery to pass and composing rhymes like: "Now I lay me down to sleep / In mud that's many fathoms deep / If I'm not here when you awake / Just hunt me up with an oyster rake."

The enemy also had plenty of time to watch the Union men inching along and trying in vain to raise horses that had fallen and drowned in the swampy roads. The Confederates taunted the

Union soldiers. Some put up signs reading THIS WAY TO RICHMOND and STUCK IN THE MUD. This was all too much. According to one account, the First Brigade of Griffin's division started tippling on their rations of whiskey, drinking a bit more than the regulation amount. They got into a battle with a division from Maine. Maine won.

Burnside tried to get back on track by ordering the men to build roads of logs, but it was very clear that the mission was not going to be salvaged. On Saturday the mucky men squished back to their original camp. Ironically, the sky cleared and the sun came out for the embarrassing retreat. After that no more major winter campaigns were attempted in Virginia.

Following the aborted battle, Burnside offered his resignation to Abraham Lincoln, who accepted it. He was replaced by Joseph Hooker, whose military reputation had earned him the nickname "Fighting Joe." The extracurricular activities of his troops added the expression "hooker" to our language.

Lest you feel too sorry for General Burnside, you should know that he did all right for himself after the war. He went on to be governor of Rhode Island and served as a U.S. Senator from 1875 until his death in 1881.

The Storm That Saved Civil War Prisoners

The Civil War took the lives of more Americans—six hundred thousand—than any other armed conflict. As with many wars, much of the suffering took place off the field of battle as soldiers starved and died of illness. Nowhere was this more true than in prisoner-of-war camps, the site of 10 percent of the Civil War deaths.

The most notorious of the Civil War camps was Camp Sumter, near Andersonville, Georgia. Built to hold nine thousand prisoners, the 16.5-acre site was chosen in 1863 because of its remote location and abundant food sources. It was located in a spot between two hillsides, with a stream called the Stockade Branch flowing in between to provide water for the inmates.

The camp was surrounded by a seventeen-foot-high stockade, and inside the stockade was another barrier, the "deadline"—the origin of the modern term. Toward the end of the war, as prison populations grew and wire for perimeter fences was scarce, authorities planted markers and painted a line between them. Anyone seen crossing the line would be summarily shot. (Is the word "summarily" ever used except in the context of being shot?) Eventually, the concept of a line of death that must not be crossed came into newspaper parlance as a time limit that cannot be exceeded.

The man in charge of the prisoners was Captain Henry Wirz. Born in Zurich in 1823, Wirz came to the United States and settled in Louisiana in 1849. At the outbreak of the War Between the States, he enlisted. In a battle in Richmond he sustained an injury to his right arm that caused him constant pain. No longer able to fight, he was sent on a diplomatic mission to Europe and upon his return was assigned to Camp Sumter. Wirz was a devout Roman Catholic but was also known for his gruff manner and use of profanity.

When the camp was first opened, it was fairly livable for the prisoners. But by the time Wirz was put in charge in April 1864, on the eve of the largest Union offensive of the war, the limited facilities had been overwhelmed, and the death toll began to climb. One of the reasons for the overcrowding was a breakdown in the prisoner-exchange program: the South refused to exchange any black soldiers they captured.

Meanwhile, sources of food were becoming scarce in the South, even outside of the prison. While hunger was a problem for both the Union and the Confederacy, it hit the South especially hard because the war broke Southern supply lines and ravaged the economy. This situation was compounded by ignorance of food preparation and storage techniques, which were seen as women's work. Thus food often arrived in soldier's camps spoiled or full of insects and rodents. Whatever food was not spoiled in transit was frequently ruined by cuisine-challenged soldiers who tried to cook over fires and turned their meals into charcoal briquettes.

Yet the soldier's meals were gourmet feasts compared to the dregs left to the prisoners. Inside the prison camps rats became a delicacy. They taste like squirrel, if Civil War diaries are to be believed. If a stray pet was foolish enough to wander across the deadline, it became a midnight snack.

By midsummer, as the war was reaching its climax, Camp Sumter packed more than thirty thousand men into the space designed for a third as many. The Stockade Branch, which provided the only water for the inmates, was backed up by the stockade's pilings. It overflowed and turned part of the camp into a swamp. To get to the stream, the prisoners had to slog through waist-high mud. When they finally arrived, they found a putrid cesspool polluted with grease from a cookhouse upstream, the waste-water of laundry, and human excrement. Those who drank the water were as likely to kill themselves with dysentery and diarrhea as to quench their thirst. About 60 percent of Camp Sumter's dead met their maker this way. Prisoners tried digging their own wells, but these too became contaminated with the wastes from the overcrowded camp.

In July, an additional ten acres were opened up, but with new prisoners being brought in at a rate of five hundred to one thousand per day it was not nearly enough. The population now numbered about thirty-two thousand, or one person per 25 sq ft (2.32 sq m). By August, when the temperature in Andersonville averages 93°F (34°C), the death toll had reached one hundred a day.

With little worldly hope, the inmates began holding prayer services asking God to send fresh water, and on August 9, he did. The skies opened up and a torrential rainstorm was unleashed on the camp. The downpour caused the Stockade Branch to overflow with such ferocity that it washed away much of the camp's foul waste. Several bolts of lightning struck near the prison, including one that hit a pine stump inside the stockade. At the base of the lightning-charred stump, a spring of fresh water emerged. The source was most likely a local spring that had been covered over during the construction of the camp, which the storm liberated. It came to be known as Providence Spring.

There was only one problem: the life-giving water was located beyond the deadline. Rising to the challenge, prisoners tied cups to poles and used them to get drinking water. Prisoners also tried digging small channels using saplings to lure the water across the deadline. Eventually, the camp officers dug a trench to bring the water in. The fact that the new spring was just out of reach probably saved many lives, as it kept the prisoners from washing in it and contaminating it.

The life-giving storm was not enough to erase Sumter's stain. By the end of the war, about thirteen thousand people had died there, making it the deadliest prison of the war, and Captain Henry Wirz became a controversial figure. He was the only person to be executed for war crimes after the Civil War and is generally remembered in the North as a cruel man who assaulted starving prisoners and allowed his guards to taunt them by sprinkling bread around the camp's perimeter and then shooting anyone who dared cross the deadline to get it. Southerners, meanwhile, tend to regard Wirz as a compassionate man who did the best he could with the resources he was given and who was made the scapegoat for the cruelty of prison life on both sides. Modern historians tend to agree with the view that the trial against him was a travesty of justice and that Wirz was made a scapegoat for all the war's sins. A memorial to Wirz erected by the United Daughters of the Confederacy of Georgia says he "became the last victim of a misdirected popular clamor . . . and [was] condemned to an ignominious death."

The site of the camp is now a memorial to prisoners of war and is known as the Anderson National Historic Site. It has been called "the most controversial site in the National Park System."

What Is That Guy in "The Scream" Afraid of?

A startled, ghostly figure stands and screams beneath a swirling, fire-red sky. Edvard Munch's painting *The Scream* is one of the most iconic images from art. It is, along with the *Mona Lisa* and Michelangelo's *David,* one of the three works of art most frequently parodied in cartoons and advertisements. Did the image spring fully formed from the nightmares of its artist or was that haunting sky a real meteorological phenomenon?

According to Munch's diary it happened this way: he was out walking with two friends when the sun began to set.

"Suddenly the sky turned blood-red. I paused, deathly tired, and leaned on a fence looking out across the flaming clouds over the blue-black fjord and towns. My friends walked on and there I stood, trembling with fear—and I sensed a great, infinite scream run through nature."

Two researchers from Texas State University believe they know exactly what caused that infinite scream. It was a volcanic eruption, which nearly destroyed an entire island, killed an estimated thirty thousand people, and altered the climate throughout the world.

Apparently the people who made the 1969 film *Krakatoa, East of Java,* did not have access to a map. Krakatoa is *west* of Java in the Sunda Strait between Java and Sumatra. It's not even properly called Krakatoa. The locals and the Dutch traders of the region both called it Krakatau, but a misspelling by a London newspaper in 1883 gave us "Krakatoa," which is what most Westerners call the place today. The origins of its name are shrouded in mystery, but one story is that a foreign captain asked a local boatman what the island was called and he said "Kaga Tau"—Malay for "I don't know." (And apparently, "I don't care.")

The island was created by volcanism caused by the convergence of the Indian and Asian tectonic plates, and prior to the great eruption in the nineteenth century, it was 2,640 feet high and featured three volcanic peaks that had been dormant for nearly three hundred years. In 1883 it was part of the Dutch East Indies. The Dutch colonizers had developed a taste for Asian spices, and Java and Sumatra had valuable sources of nutmeg, cloves, and pepper. With their spice money the Dutch created tree-lined streets, canals, and elegant entertainment venues like the Concordia Club.

When the mountains of Krakatau began to rumble in August 1883, the residents of Java paid little attention. It was a pleasant summer, and families enjoyed days on the beach as the circus came to town. Only a circus elephant had the good sense to try to get out—he went berserk in a hotel room, much to the confusion of his handlers.

On Sunday, August 26, Krakatau became impossible to ignore. Clouds of smoke spewed into the air, blotting out the sun. Hot ash rained down. There were three explosions of increasing size. Then there was the big one: Krakatau erupted with a force of ten thousand Hiroshima bombs. The deafening sound could be heard 4,600 km (2,858 mi) away. The blast set off a series of tsunamis that killed nearly thirty-seven thousand people on the islands of Java and Sumatra; 160 villages were washed away. High waves were felt as far away as South Africa, India, Japan, and Australia. A Dutch warship was picked up and deposited two and a half miles inland on a hillside, where it remained for many years. More than six cubic miles of rock were hurled upward. Dust and ash blanketed the region and rose into the atmosphere, altering the weather—temperatures across the globe dropped and did not return to normal for years.

The eruption may even have created a new form of cloud. Noctilucent clouds, those wispy clouds in the upper atmosphere, were first spotted by German meteorologists in 1883. They believed that Krakatau seeded the mesosphere with dust, creating this new formation. Not all scientists are convinced that the volcano caused them, since they still exist today, and the dust should have dissipated by now.

Dust and ash in the stratosphere interfered with sunlight and created dramatic sunsets as far away as Northern Europe and the United States. A team of firemen from Poughkeepsie, New York, even rushed to put out what they thought was a fire, only to discover it was a fiery sunset.

The Krakatau sunsets appeared as blue, purple, pink, bronze, or brown and sometimes appeared as colorful rings around the sun. The vivid colors inspired artists and poets, including Edvard Munch. The British artist William Ashcroft created watercolors of the colorful skies, as did the Hudson River School painter Frederic Erwin Church, and Alfred Lord Tennyson was probably thinking of the Krakatau explosion when he wrote in "St. Telemachus:" "Had the fierce ashes of some fiery peak / Been hurl'd so high they ranged about the globe? / For day by day, thro' many a blood-red eye / The Wrathful sunset glared."

The Krakatau effects reached the skies of Norway in late November 1883 and lasted until mid-February 1884. The most famous version of Edvard Munch's *The Scream* was painted in 1893 as part of *The Frieze of Life*, a group of works with themes of sickness, anxiety, and alienation, that drew on the artist's experiences, including the death of his sister. The aggression of his work was controversial at the time, sometimes branded "obscene" by critics.

The image of *The Scream* that we are most familiar with was one of four. The earliest sketch was from a different angle—most likely the spot where Munch actually saw the sunset. A historical marker on a road called Valhallveien in Oslo commemorates this moment.

When the Texas State University researchers traveled to Oslo to look at the spot, they found the marker was at a horseshoe bend in the road that didn't exist in the 1800s. They did some further investigation and determined that the actual spot must have been on a road now called Mosseveien. The view of the fjord matches the original sketch. In nineteenth-century photographs, that stretch of road was shown to be lined with railings like those in *The Scream*. When the researchers went to that spot, they could see that Munch was looking to the southwest—exactly where the Krakatau sunsets would have appeared.

Some believe the Krakatau sunsets, or a similar effect produced by another storm, may have been responsible for the visions of pilgrims who came to a field in Fátima, Portugal, after three shepherd children claimed to have seen apparitions of the Virgin Mary there in 1917. After the children reported what they had witnessed, a crowd of seventy thousand saw blue, silver, yellow, and white circling the sun.

"Similar sorts of coloured suns are sometimes seen after desert duststorms, and it is possible that the crowds at Fátima witnessed a cloud of dust blown over from the Sahara," wrote weather reporter Paul Simons in *The Times* of London. "But why such a phenomenon would occur at just the right time in front of such a large crowd is a theological, rather than meteorological, question."

A Gust of Wind and Aviation Obscurity

What aviation pioneer was the first to fly in a heavier-than-air craft? Had it not been for a storm and a poorly timed gust of wind, the answer might have been Samuel P. Langley.

In the early 1900s, Langley was America's leading aeronaut. He was the secretary of the Smithsonian Institution in Washington and literally wrote the book on the theory of mechanical flight. Prior to that he worked at the Harvard and Allegheny observatories and the U.S. Naval Academy. He counted among his friends some of the most important men in the nation. (The Langley Air Force Base, the oldest continuously active base in the United States, was named in Samuel P. Langley's honor.)

With his contacts and reputation, the would-be aviator was able to obtain funding for his experiments from the Board of Ordnance and Fortification of the War Department. Officials believed that an "aerodrome" could be used by the military to observe troop movements. After the outbreak of the Spanish-American War they granted Langley $50,000 for "developing, constructing, and testing a flying machine capable of bearing a man."

Up to this point, Langley had constructed several working models powered by steam and later by gasoline. Each model yielded valuable information about the lift and drag of various surfaces. Finally he designed a gasoline-powered, four-winged machine that was said to resemble a giant dragonfly. His biggest problem, however, was the weight of the engine. At that time gasoline engines were far too heavy to get off the ground. So Langley designed his own light engine with a flash-boiler steam system. In 1896, he tested an unmanned model, and it flew three thousand feet. Later that year, an updated model flew forty-two hundred feet, but a machine that could carry a man was still far from being a reality. A year and a half behind schedule on his promise to develop manned flight, Langley ran out of his government money, but the Smithsonian Institution pitched in with the $23,000 he needed to complete the project.

Finally, in July 1903, Langley believed he had overcome his technical problems and he towed his perfected machine to the shore of the Potomac River to launch it. The aerodrome could not take off on its own power; it had to be catapulted from a barge in the river. Thanks to an ill-timed storm, the aerodrome would not fly into history that day. When the clouds cleared, the aerodrome's wings had been warped to the point that it would not be airworthy. Langley dragged it back to his shop for three months of repairs.

On December 8, 1903, Langley once again attempted to become the first man to demonstrate a working heavier-than-air craft. The press, military observers, and members of Congress lined the shore to witness the historic event. The machine was placed on a houseboat, pulled into the Potomac, and faced directly into the wind. At 4:45 a pilot named Charles Manley signaled the crewmen to release restraining pins so the plane would be thrown into the wind by a spring-driven catapult. But just as the pin was pulled, a heavy gust of wind sent the platform lurching. The aerodrome's rear wings collapsed, and it made a spectacular nosedive into the water.

The aerodrome became a punch line and Langley was widely ridiculed for building it. The *New York Times,* for example, wrote, "We hope that Professor Langley will not put his substantial greatness as a scientist in further peril by continuing to waste his time and the money involved in further airship experiments. Life is short, and he is capable of services to humanity incomparably greater than can be expected to result from trying to fly . . . for students and investigators of the Langley type, there are more useful employments."

This was probably one reason why not a single reporter showed up to watch the Wright brothers successfully fly the world's first airplane, the *Flyer,* nine days later. In fact, only one newspaper, the *Virginian-Pilot,* gave the Wright brothers any next-day coverage. Their story was pieced together from witness accounts.

Whether the wind and the catapult were entirely to blame for Langley's failure is a matter of speculation and controversy. Shortly after Langley's death in 1906, his successor at the Smithsonian Institution launched a campaign to give credit for manned flight to his mentor. The Smithsonian allowed Glenn Curtiss to restore and fly the original aerodrome. He was hardly a disinterested scientist—at the time he was being sued for patent infringement by the Wright Company and probably thought proving Langley flew first would bolster his legal claim. So Curtiss made a few modifications to be sure the aerodrome would get off the ground. He argued that these were only replacements of broken parts, but others believe they were significant modifications. In any case, Curtiss got rid of the houseboat-based launching rail and replaced it with pontoons; and the machine flew. Because of the Smithsonian's efforts to downplay the Wrights' feat, the original 1903 *Flyer* was lost to the Smithsonian Institution and the United States. It is now on display in the Science Museum in London.

El Niño and Dashed Polar Dreams

Imagine trekking ever-onward, braving frostbite and exhaustion in the quest to be the first to reach the South Pole, only to arrive and find someone else's flag and a letter congratulating you on being second. This was the fate of Robert Falcon Scott and his team of explorers. But it gets worse: having lost the race to the Pole, they would also lose their lives in the attempt to get home.

Captain Scott, the leader of this ill-fated expedition, has been recorded in history as an ill-prepared bungler who brought about his own demise through bad leadership and poor planning. Modern meteorologists, however, blame it on the weather.

Dr. Susan Solomon, a scientist at the National Oceanic and Atmospheric Administration, analyzed seventeen years' worth of meteorological data from stations in Antarctica and compared it to the information from the diaries of the men on the Scott expedition. Scott had the misfortune to be making his trek home during a freak cold spell, with temperatures nearly 20°F (–6.67°C) less than the average for the season.

In the early twentieth century, the race to conquer the last unexplored continent was as charged as the space race between the United States and the Soviet Union would be in the 1960s. Japan, Germany, Sweden, and Belgium all set their sights on the South Pole, but the main contenders for the First-on-the-Pole prize were the United Kingdom and Norway. Captain Scott's rivals in the field were Ernest Shackleton, who had come within one hundred miles of his goal on January 9, 1909, and the Norwegian Roald Amundsen. Amundsen was one of his country's most celebrated explorers. He had already sailed through the Northwest Passage and was one of the first to spend a winter south of the Antarctic Circle.

In 1910 Scott was raising funds for a push south when he received a telegram from his Norwegian rival. "Beg leave to inform you [expedition] proceeding Antarctic. Amundsen." If the British were going to beat Norway to the Pole, they would have to step up their plans. Scott quickly assembled a team of twenty-four men, nineteen ponies, thirty-three dogs, and innovative motor sledges.

In addition to their slight head start, the Norwegians had certain natural advantages over the English. They came from a polar country. Amundsen, legend has it, slept with his bedroom window open when he was a boy, even on the coldest Norwegian nights. "Scott's party had six years of [skiing] experience between the five of them, whereas Amundsen's group had a century of skiing between them," historian Roland Huntford told the BBC.

The British and Norwegian teams both arrived on the southern continent in January 1911 and set up camp. Scott's team planned to follow the route Ernest Shackleton had used on his push toward the Pole. Amundsen plotted an alternative—and superior—route.

This is not to say that all was fun and games in the Norwegian camp. On their first attempt toward the Pole they encountered temperatures of –40°F (–40°C). They were forced back into their camp. This caused one of the men, the famous explorer Hjalmer Johansen, to mutiny. Amundsen dismissed three of the men and set off on October 20 with a team of five men (including himself): Oscar Wisting, the skiing champion; Olav Bjaaland; and two expert dog-drivers, Halmer Hanssen and Sverre Hassel. When Amundsen's team became national heroes, the news of Johansen's rebellion put an end to his career. He fell into a depression and took his own life.

Captain Scott began his southern journey on November 1, 1911. Despite the later start, Scott expected to make up the difference by using his motor sledges. Unfortunately, they did not operate well in the cold and were abandoned. Next, the ponies began to fall. This slowed the expedition even further. As the dogs began to suffer, Scott ordered them back to base camp.

At 150 mi (240 km) from their goal, the last support team turned back to the camp. For reasons unknown, Captain Scott made a last-minute decision to add a fifth man to the final party. Originally, only four men were to make the last leg of the journey, but Scott decided to take Henry Bowers along. The supplies, however, had been figured for a team of four. They could be stretched for the extra man, but this left little room for error.

On January 17 Scott and his hungry, weary team finally arrived at the Pole only to find it decorated with Amundsen's Norwegian flag. The Norwegians had stepped foot on that soil on December 15, when the temperature was eight degrees warmer. They were already on their way home, having traveled back through comparatively mild daily minimum temperatures of around 5°F (–15°C)

"Great God!" Scott wrote in his diary. "This is an awful place and terrible enough for us to have laboured to it without the reward of priority."

To make things worse, the dispirited British expedition was

about to be attacked by El Niño. This weather system affected the Antarctic weather, causing a drop in temperatures to –40, about 8°C (15°F) lower than average. Their trip away from the Pole was colder than it had been on the Pole itself (around –20°C/–20°F). The temperatures stayed atypically low throughout the team's return trek. In the past fifteen years of modern measurements, only one year's temperatures were comparable to those experienced by the Scott expedition.

The temperature drop affected more than the comfort of the explorers. At such a low temperature, the film of water that enables runners to glide did not form. It was like pulling a sledge over grit. The team hoped to literally sail home, attaching sails to the sleds and using wind power to guide them. Unfortunately, when the temperature dropped, the winds became still. The exhausted, frostbitten men could travel only five to eight miles a day instead of the fifteen to twenty miles a day they had projected.

"On this surface we know we cannot equal half our old marches and that for that effort we expend nearly double the energy," Scott wrote.

The first man to succumb was Petty Officer Evans, who died on February 17. A month later, Captain Oates, knowing that he was becoming a burden to his comrades, sacrificed himself. Crippled with frostbite, he walked out of the tent with the announcement, "I am just going outside, and may be some time."

"We knew Oates was walking to his death," Scott wrote in his diary. "It was the act of a brave man, and an English gentleman."

The remaining three struggled on for ten miles. They made it to within eleven miles of the "three degree" depot, where they had left stores of food and fuel on the journey down. They could go no farther. Their camp was hit by a blizzard that lasted for more than a week. The men were slowly starving, suffering from scurvy, hypothermia, and exhaustion. They were in no condition to travel in a blizzard. All they could do was stay in the tent and await their inevitable death.

"Our wreck is certainly due to this sudden advent of severe weather, which does not seem to have any satisfactory cause," Scott wrote. "We shall stick it out to the end, but we are getting weaker, of course, and the end cannot be far. It seems a pity but I do not think I can write more." It was his last journal entry. It was found on November 12, 1912, in the tent with the frozen bodies of the explorers.

Cold Shaving and Jacob Schick

Necessity is the mother of invention, and sometimes weather is the mother of necessity. Imagine for a moment that it is 1910 and you are in the wilds of Alaska. The region, not yet a state, has been a U.S. territory for only a little more than forty years. It is a frigid frontier, a wilderness inhabited by prospectors, soldiers, fur traders, and a few hearty missionaries.

You have made camp on the frozen ground, but your humble shack does little to protect you from the howling winds and the –40°F (–40°C) temperatures. You fill a basin with water, and the bowl develops a crust of icy frost. Into this chilly liquid you plunge a safety razor. As you stare into the mirror, anticipating the sting of the nearly frozen steel on your face, you wince.

This was the exact situation that Colonel Jacob Schick found himself in one Alaskan morning. Schick, a native of the much more temperate Ottumwa, Iowa, had served in the Philippines during the Spanish-American War. The conditions didn't agree with him, and after he suffered a severe attack of dysentery, a doctor suggested he be transferred farther north for health purposes.

The army took this suggestion quite literally, and Schick soon found himself in Alaska, where he helped lay out military telegraph lines across one thousand miles of the Alaskan interior. Alaska, at this time, was in the throes of the Yukon gold rush. More than one hundred thousand people sought their fortunes in the inhospitable landscape, but fewer than thirty thousand ultimately survived. Schick had become reasonably acclimatized to the harsh surroundings, and when he retired from the army, he decided to peek around for some gold himself.

He didn't find any, but he did come up with an idea that was as good as gold. Prospecting in the dead of winter, Colonel Schick sprained an ankle and was forced to remain close to camp while he recuperated. His cold-weather camp was a great test of his survival skills. Schick killed a moose for food, drew water from a stream that was iced over, and cooked over an open fire. Whereas a lesser man might simply have grown a beard, Schick continued his warm-weather grooming habits. Each pass of the razor over his stinging skin increased his motivation to create a more perfect shaver.

Shaving, incidentally, is an ancient tradition. Personal grooming experts say it dates back to the Paleolithic era. Shaving materials have been uncovered at archaeological sites at least seven thousand years old. There are clean-shaven men in prehistoric cave drawings. Alexander the Great went beardless for reasons of vanity and military security (beards make a great handle if you plan to whack off somebody's head).

Razors of those BC times were not too different from kitchen knives. In 1927 the editor of *Antiques* magazine called straight razors "the lethal weapons with which the men of a courageous generation were wont to keep the verdure of their open and unprotected countenances in a state of decorous subjugation." In other words, manipulating an open blade close to your jugular vein can have serious consequences.

Still, men, the courageous and secretly vain creatures that they are, were willing to take the chance and over the centuries the presence or absence of a beard took on a variety of different meanings in different cultures. In Rome, for example, the wealthy were clean-shaven while the slaves and the masses kept beards. The word "barbarian," in fact, means "bearded one" and it was how the naked-faced Romans referred to their bearded foreign invaders. In Turkey, on the other hand, slaves were forced to shave, while the wealthy grew their facial hair.

The comfort of shaving took a great leap forward in 1903 when King Camp Gillette invented his disposable razor. Gillette was a salesman for a bottle cap company and his employer, who had invented the cork-lined bottle cap, suggested Gillette invent something as he had. Something "which once used is thrown away, and the customer keeps coming back for more."

Gillette found his inspiration in the morning mirror, while shaving. After trying 700 different blades and 51 razors, Gillette had his prototype. His razor was inexpensive, it allowed men without martial-arts training to shave themselves, and best of all it had blades that wore out and couldn't be sharpened. After only two years, Gillette had sold 250,000 razors and 100,000 blade packages and had single-handedly changed the way American men shaved.

Still, Gillette's disposables had their drawbacks. The "safety" razor wasn't really all that safe. To change a dull blade, you had to take the thing apart. This often led to cuts on the fingers and hands. And then there was the need for water, the bane of the Alaskan shaver's existence.

Which leads us back to Jacob Schick. During recuperation, he had lots of time to ponder the mysteries of beard removal. He drew up a blueprint for a device that would need no lather or water. Inspired by the repeating rifle, he called his design the

Magazine Repeating Razor. He sent his plans off to a number of manufacturers, all of whom rejected them.

Schick's business ambitions were put on hold when World War I broke out. He returned to active military duty as a captain, served in England, and eventually retired as a lieutenant colonel in 1919. He continued to believe, however, that the dry shaver was his greatest accomplishment. He and his wife, Florence, mortgaged the family home at one point to keep the shaver dream alive.

The main problem with his electric razor was that he had not yet perfected a motor that was powerful enough yet small enough to be practical. That breakthrough finally came in 1928. The perfected dry shaver went on the market in 1929. The razor was touted as "one of the greatest mechanical inventions of the twentieth century." *Collier's* magazine wrote, "Shaving as it is practiced in America today marks a step forward in civilization." And it was all because of the dastardly low temperatures in Alaska in winter.

The Failure of Forecasting and the Death of Lord Kitchener

On June 5, 1916, Lord Horatio Herbert Kitchener drowned off the coast of the Orkney Islands. Kitchener, the hero of Khartoum, was a celebrated figure in England. As Secretary of State for War during the Great War (World War I), he expanded the army from twenty divisions in 1914 to seventy in 1916. His very image became a call to arms, a symbol of the British determination to win. It was his serious, mustached face pointing out from the recruitment posters which announced "Your Country Needs You!" Those who were alive that day would always remember exactly where they were when they heard the news of Kitchener's death. A nation plunged into a state of collective grief and mourning similar to the feelings evoked in more modern times by the death of Princess Diana. That such a man could die so suddenly and senselessly was incomprehensible. All kinds of conspiracy theories flew about. Some refused to accept that he was dead and believed the stories were invented to fool the Germans. The truth was far too simple. Kitchener was a victim of poor judgment and foul weather.

The ambitious Kitchener began his military career in the Royal Engineers. In 1886 he was appointed governor of the British Red Sea territories and eventually became commander in chief of the Egyptian army. In 1898 he crushed the separatist Sudanese forces in the Battle of Omdurman and then occupied the city of Khartoum. This victory sealed his reputation. Following this, he served as governor of Sudan, commander in chief of the Boer War, then commander in chief in India and in September 1911, proconsul of Egypt. He stayed there until war broke out in Europe in 1914, which is when he was named Secretary of State for War. He was virtually alone in his belief that the war would drag on for some time, and he rapidly enlisted and trained new recruits, who came to be called Kitchener armies. They went off to battle with a letter from Kitchener in their Active Service Pay Book.

> You are ordered abroad as soldier of the King to help our French comrades against the invasion of a common enemy. You have to perform a task which will need your courage, your energy, your patience. Remember that the honour of the British Army depends on your individual conduct. It will be your duty not only to set an example of discipline and perfect steadiness under fire but also to maintain the most friendly relations with those whom you are helping in this struggle. The operations in which you are engaged will, for the most part, take place in a friendly country, and you can do your own country no better service than in showing yourself in France and Belgium in the true character of a British soldier. Be invariably courteous, considerate and kind. Never do anything likely to injure or destroy property, and always look upon looting as a disgraceful act. You are sure to meet a welcome and be trusted; your conduct must justify that welcome and trust. Your duty cannot be done unless your health is sound. So keep constantly on your guard against any excesses. In this new

> experience you may find temptations both in wine and women. You must entirely resist both temptations, and, while treating all women with perfect courtesy, you should avoid any intimacy.

While the public idolized him, the view was not shared by the members of the cabinet. Kitchener's distaste for teamwork led to conflicts. In 1915 he was blamed for a shortage of shells, and he was stripped of that responsibility; and later that year he lost control over strategy. He offered to resign, but the government feared the effect this would have on morale among a public that still adored him.

Writing in 1925, the meteorologist and author Alexander McAdie summed up the public sentiment: "He had appeared before a Committee in the House of Commons and given such a clean-cut, concise statement of what had been accomplished under his stewardship as Secretary of State for War that those who came to scoff remained to pray. Those who would have forced him from his high office—the politically hungry, the ever present critics, instead of censuring could only compliment."

Fresh from this political battle, it was a tired and dispirited Kitchener who embarked for Russia on June 5, 1916. Kitchener was headed to the nation at the personal invitation of Czar Nicholas to drum up support for the war effort. A strong northeast gale was blowing across stormy seas and the admiral gathered his staff to discuss the best route for Kitchener's ship, the *Hampshire.* The *Hampshire* was an armored cruiser that was to convey Kitchener around the north cape of Norway into the Russian port of Archangel. After some discussion, they agreed that it should take a western channel and then head north in the lee of the Orkneys, where it would be sheltered from the worst of the weather. Minesweepers had not yet been through the channel, but merchant ships had been using the lanes for months without incident.

Unbeknownst to them, the German U-boat *U-75* had laid a mine in the passage only a few days before.

No sooner had they set sail than the storm center passed, and the wind changed direction. A fierce northwest wind raised dangerous waves. The best course of action would have been to wait the storm out, rather than try to outrun it, but this is not what they chose to do.

After a couple of hours the two torpedo destroyers that escorted the *Hampshire* encountered a head-on gale. Unable to keep pace, they turned back. The *Hampshire* continued on. The seas were so high that minesweeping operations were not possible. Around 7:40 P.M. the steamer, a mile and a half from the Brough of Birsay, struck a submerged mine. Onlookers saw Kitchener and his aide, Oswald Fitzgerald, on the bridge as the bow went under, and the ship sank. By eight o'clock, the *Hampshire* was at the bottom of the sea with all but twelve members of its crew.

The idea that the war minister could be killed on such a mission, only two miles from the naval stronghold of Scapa Flow, was inconceivable. An army volunteer, interviewed by the *Sunday Telegraph* at age 101, still remembered his reaction to the news of the war minister's demise. "Oh my God," he said. "What's going to happen to us now Kitchener's gone? He was the driving force in this country. We thought we'd probably lose the war as a result."

Some, like the journalist Frank Power, played on the public's denial. Power, whose real name was Arthur Vectis Freeman, was a failed filmmaker who took up newspaper storytelling when his cinemas went bankrupt. He reported that Kitchener had not really been aboard the ship; he had been impersonated by a double. The real man was alive and well in Norway. As time passed and Kitchener never commented, his readers grew skeptical but no less fascinated by reports of their hero. So the story changed. Kitchener, Power explained, had been killed by his enemies within the navy.

He was buried in an unmarked grave in Norway, and Power was going to go and retrieve his body.

Hope was so great that Power was able to drum up support from eighty-nine branches of the British Legion. In 1926, as they called for an investigation of Kitchener's death, Power sailed to Norway to collect his remains with a film cameraman in tow. The cameraman recorded the procession as a flag-draped coffin was transported to a chapel. Teary-eyed mourners lined the route to pay their last respects to a man who had embodied England's hopes and dreams. The police were out in full force to provide security.

There was one thing that Power had failed to take into account. For a person to be buried in England, he must be examined by a coroner and be given a death certificate. The pathologist, Sir Bernard Spilsbury, opened the coffin, but it would not have taken an expert to recognize that it was not Kitchener lying there. Instead, the box was full of tar in a quantity sufficient to simulate a man's weight. Power, who had a conspiracy theory for every occasion, claimed that the government had stolen the body, but no one was buying it.

This time the police investigated Power. They learned that one part of his story was true. He had discovered Lord Kitchener's coffin. It had been made in case his body was ever discovered, but it sat empty in a mortuary for a decade until Power bought it for £12. Power had never been to Norway. He had seen the whole exercise as an elaborate movie and PR campaign. The resulting film, *Kitchener's Coffin,* was banned in England, but Power was not prosecuted for the fraud. The story does not end there, though. A man named Peter Gates brought a civil lawsuit against the filmmaker. His beef? He said Power had stolen the idea of the hoax from him. A judge threw his case out of court.

After Kitchener's untimely death, a city in Ontario, Canada, changed its name from the unfortunately German (during the Great War) "Berlin" to "Kitchener." You'll find Kitchener in the Grand River valley. Today, because of its close ties to nearby Waterloo, the region is commonly known as Kitchener-Waterloo.

Did the death of this one man forever change the course of history? Probably not, but his contemporaries certainly thought so. Alexander McAdie expressed the general mood:

> All admit that K. of K. was the one man in Europe who might have held the wavering irresolute Ruler of all the Russians to a fixed purpose. With the earnest support of the Grand Duke Michael, needed reforms might have been accomplished and the ominous rumblings of impending revolution silenced. He might have brought to the distressed nation (and probably no one else could have done so) authority that would be respected, integrity and disinterestedness that all factions would have appreciated. If the storm center had passed over the Orkneys a few hours earlier, the eastern channel would have been selected. If the fury of the northwest wind had been less, a rescue would have been effected. But it was not to be. The destiny of Russia, perhaps the fate of Europe itself, hung upon a forecast of weather made that June afternoon in Scapa Flow.

Rain Clouds Put an End to the Age of the Airship

While much is made today of the space race, little is said about an earlier technological competition: the airship race. Back in the 1920s and 1930s, dirigibles looked like the transportation of the future. Unlike airplanes, dirigibles were quiet, they had lower fuel costs, and the interiors of the finest passenger craft were roomy and luxurious. Germany and the Soviet Union took part in the race to produce the largest, most impressive dirigibles as demonstrations of superior technology. Yet the era of the airship ended abruptly on May 6, 1937, when the *Hindenburg* burst into flames during a landing at Lakehurst, New Jersey. For years it was assumed that the craft was destroyed by a hydrogen explosion. More recent research suggests that the *Hindenburg* was a victim of ordinary rain clouds.

The airship era began on July 2, 1900, three years before the first airplane flight. On that date Count Ferdinand von Zeppelin, a retired German brigadier general, took the first rigid airship on an eighteen-minute flight. In 1910 the rigid *Deutschland* became the first commercial dirigible, and between 1910 and the beginning of

World War I, German airships flew 107,208 miles and carried more than 34,000 passengers without any injuries.

It would take some time for Zeppelin and his backers to convince the German military that the dirigible would have military uses. When they did catch on, they embraced it fully. While the British, French, Italian, and U.S. militaries also operated blimps—that is, nonrigid aircraft—during the Great War, the Germans created a fleet of military dirigibles that were the envy of the world. Writing in 1917, Frank H. Simonds, author of *History of the World War*, described the enthusiasm for the craft.

"Planes have proved powerless to thwart the Zeppelin," he wrote. "How keenly the British authorities regret that England did not build a fleet of Zeppelins ten years ago. . . . They gave the Germans an enormous advantage at the outset. . . . As a bomb-dropper, the Zeppelin is more successful than the airplane. Its sighting instruments are not only more elaborate and accurate, but manipulated in ease and comfort."

They were so successful that many British citizens called them "baby killers" for their role in bombing English cities. They were considered to be such a threat that German airships were banned under the Treaty of Versailles. During this brief hiatus, other nations scrambled to catch up. The British used a captured zeppelin as the model for two of their own airships, which became the first aircraft to make a round-trip crossing of the Atlantic.

The restrictions on German airship building threatened to destroy the Zeppelin company. When a British-made dirigible constructed for the U.S. Navy crashed, Hugo Eckener, then the head of the Zeppelin company, was able to convince the United States to loosen restrictions and let the masters take a stab at giving them the best airship afloat. They called it the *Los Angeles*.

Its success inaugurated a new era of transatlantic travel. The follow-up to the *Los Angeles* was the famous *Graf Zeppelin*, which became a symbol of luxurious and futuristic travel. The *Graf Zeppelin* completed the first transatlantic passenger flight and later the first round-the-world flight in August 1929. When the zeppelin passed over Russia on this tour, it created quite a stir. The Soviet Union immediately set to work on its own airship program. The government came up with a five-year plan, which began in January 1933. The goal was to build a fleet of fifty ships.

The pride of the Russian airship fleet, the *V6,* made its maiden voyage on November 5, 1934, as part of the celebration of the anniversary of the October Revolution. It was supposed to be a passenger ship, but there was one little snag: There were no hangars big enough to hold it. Instead, the dirigible became a symbol of Russian pride, as it had been a symbol of German pride. The *V6* flew over Soviet cities to tout the Communist Youth Organization's twentieth anniversary. (The Russian film *Burnt by the Sun* portrays such a campaign.)

Germany's newly powerful National Socialist Party, the Nazis, also took full advantage of the visual effect of the airship. The *Graf Zeppelin,* with swastikas on its fins, flew over Germany, dropping pamphlets touting Nazi ideology.

The Zeppelin company was preparing to unveil its greatest accomplishment yet, the largest man-made object to fly. The elegant *Hindenburg* would dwarf everything in the sky. Although Hugo Eckener himself was anti-Nazi, the use of Zeppelin craft as a Nazi propaganda tool and fear of the airship's potential use in a future war led to the American passage of the Helium Control Act. The United States was the only large-scale producer of helium at the time and the act was designed to keep it out of German ships.

Therefore the *Hindenburg* had to be filled with a more volatile, flammable gas: hydrogen. But more crucial to the craft's fate would be the choice of a coating for its skin. It was made of iron oxide covered with cellulose acetate, which was designed to protect it from moisture. The highly flammable mixture was practically identical to rocket fuel. As if to ensure it would burn, the paint that covered the acetate was stiffened with powdered aluminum, which is also highly combustible.

The *Hindenburg* was launched in Friedrichshafen, Germany, in April 1936. After nearly a dozen transatlantic flights, the public was still fascinated with the magnificent silver craft. As it came in for a landing in Lakehurst, New Jersey, on May 6, 1937, a group of newsreel cameramen and radio reporters were on hand to record the event.

Yet it was to be quite different than they had imagined, thanks to rainy weather. The ship, which was already delayed due to headwinds over Newfoundland, was unable to dock because of the stormy weather. It circled the airport for more than an hour waiting for the weather to clear. As the *Hindenburg* passed through rain clouds, the craft became negatively charged. When the crew dropped the wet lines to dock, they acted as a ground. When the metal frame of the ship earthed its charge, the skin heated up and the highly flammable coating ignited. Within ten seconds most of the ship was ablaze; by the time thirty-four seconds had passed, the mighty *Hindenburg* was a burning mass on the ground.

What made the turn of events such a blow to the airship industry was the presence of reporters. The entire tragedy was captured on film and broadcast live on the radio.

"This is one of the worst catastrophes in the world," radio reporter Herb Morrison famously observed. "Oh! The humanity and all the passengers!"

The dramatic image of the *Hindenburg* engulfed in flames

threw the safety of the entire industry into question. The Zeppelin company, which had been building another *Hindenburg*-size craft at the time of the accident, was out of business by 1940. The U.S. Navy's Lighter-Than-Air program, begun with promise in 1921, was defunct by 1961.

Gee, It's Cold in Finland
The Winter War

It seems that every war that involves Russia has an element of freezing to death. Generally, this has worked to the Russians' advantage. In 1939, however, the tables were turned. The Red Army, under the command of General Kirill Meretskov, was about to march confidently into Finland. What the military could not know was that the winter of 1939 would be the coldest the region had seen since 1828. What was supposed to be a "walkover" lasting only a couple of weeks would become a long and brutal descent into a frozen hell.

The foundations of this debacle were laid in August 1939, when the Soviet Union and the Third Reich signed a nonaggression pact. Josef Stalin was concerned about German military expansion and needed time to rebuild the Red Army, which he personally had decimated with a purge of the officer corps in 1937. Hitler wanted the pact so that he could invade Poland and fight France and Britain without any interference from the Soviets.

The pact, which was supposed to last ten years, said that neither power would attack the other. Along with the public pact, the two nations signed a secret protocol, which divided Eastern Europe into German and Soviet regions. This protocol gave Lithuania, Latvia, Estonia, and Finland to the "Soviet sphere of influence." The problem was, no one bothered to tell the Finns. As it happened, they had their own opinion on the matter.

The task of making Finland aware that it was to take orders from Moscow fell to General Meretskov. The general had good reason to feel confident about his mission. The entire eight hundred-mile Russo-Finnish border was lightly defended. The only real challenge was at the Mannerheim Line, which was guarded by nine poorly equipped Finnish divisions. Initially, even the weather seemed to be working in the Soviets' favor. The lakes and rivers were frozen solid enough for the Red Army to storm across them, but there was little snow to block the roads.

At 8:00 A.M. on November 30, the order came for the attack on Finland. The Red infantry, under cover of an artillery bombardment, charged across the Finnish border, firing machine guns and screaming the battle cry "Urra!" An estimated six hundred thousand Soviet troops rolled across the Finnish border. Soviet propaganda had painted a picture of a Finnish population that longed for a sensible Communist government. The soldiers, convinced they were Finland's liberators, crossed the border with gifts, clothing, money, and Finnish-language leaflets. If, for some reason, the Finns were not amenable to liberation, the soldiers had no choice but to march on, for behind them were *Politruks*, political commissars, who would shoot anyone who turned back.

In contrast, the Finnish army was a mishmash of reservists with little training. Most wore their own civilian clothes. They addressed one another by first name. They sometimes saluted, but they were not particularly conscious of status or rank. Given the vast differences between the Russian and Finnish armies, Finnish field marshal Carl Gustaf Mannerheim decided that their only chance was to spread the troops out as far as he could. They would attack in small numbers, without artillery cover, and use the element of surprise, gliding by on skis at night or during fog and snowstorms.

What they lacked in tanks and armaments, the Finns made up for with grit, imagination, and winter survival skills. In Finland they say that one soldier, watching the Russian display of overwhelming force quipped, "So many Russians. Where will we bury them all?"

While the Soviets marched in dark uniforms that all but flashed neon against the white background, the Finns wore white and attacked on homemade skis with their Russian-designed machine guns hanging from their sides on leather belts.

Still, one man on skis is no match for a tank. The solution to this problem came from the *Alkohooliliike*, the State Liquor Board. They supplied forty thousand liquor bottles, which the Finnish fighters filled with a mixture of kerosene, tar, and gasoline. They inserted a rag in the bottle, ignited it, and threw it at the back of a Soviet tank, each of which had an extra fifty-gallon gas tank on the back end. Excess oil on the compartments quickly burst into flame. More than two thousand tanks were destroyed with these homemade bombs. The Finns named them "Molotov Cocktails" after the secretary of the Central Committee of the Communist Party, Vyacheslav Molotov. (Molotov, known for his cold aspect, was once described by a British diplomat as "a refrigerator when the light has gone out." Winston Churchill said he had the "smile of Siberian winter.")

It would be six months before the Red Army caught on to the idea of camouflaging their tanks for snow. They also tried to beat the Finns at their own game, and a new manual was sent to the Russian troops on how to fight with bayonets on skis. The manuals were worthless: it is physically impossible for a man on skies to use a bayonet because the energy of the thrust forces the attacker to slide backward.

As the Russians learned these painful lessons, the temperatures in Finland began to drop. The *Talvisota*, or Winter War, brought a –40° to –50°F (–40° to –45°C) deep freeze to the battlefield. The chill affected soldiers on both sides, but the Finns were on their home turf. They were accustomed to this weather, and they had more access to warm clothing than the invaders had. They dressed in layers and wrapped up in bedsheets. Their families, friends, and supporters sent reinforcements—gloves, socks, sweaters, scarves. The Russians, meanwhile, were dressed in the light clothing they had put on for their two-week trek. Many had no overcoats and no gloves—only mittens, which they had to take off in order to fire their weapons. They huddled around fires near their useless, frozen tanks, battling frostbite, waiting to freeze to death.

Meanwhile, the comparably warm Finns continued their surprise attacks on skis. Often they targeted the Russian field kitchens, knowing that the combination of hunger and cold could mean the death of an entire unit. Virginia Cowles, a reporter for the *Sunday Times* and *New York Herald Tribune* described the carnage. She saw hundreds of corpses frozen like statues in the positions in which they died. "I saw one with his hands clasped to a wound in his stomach; another struggling to open the collar of his coat, and a third pathetically clasping a cheap landscape drawing, done in bright childish colors, which had probably been a prized possession that he had tried to save when he fled into the woods."

When the Russian soldiers complained about their plight, the *Politruks* noted their disloyal comments for the record. Some men did manage to write letters home, but most of the letters were never delivered. They were found on their dead bodies, clutched in their frozen fingers. "Why were we led to fight this country?" read one of these messages.

In the end, despite the horrible losses suffered by the Russians, Finland was no match for the power of the USSR. Air bombardments on the Karelian Isthmus overpowered Finnish resistance, but it was a Pyrrhic victory for Russia. According to some estimates, of the 1.5 million men who were sent into Finland, one million would never return.

On March 13, 1940, Moscow and Helsinki signed an armistice. Under its terms, Russia received twenty-two thousand square miles of formerly Finnish territory, including the Karelian Isthmus, Finland's second-largest city, Viipuri, and the shores of Lake Ladoga. It was, as one Russian general put it, "enough ground to bury our dead."

Although the Winter War is often neglected in history texts outside Finland, it had far-reaching consequences. Stalin learned from the fiasco and reorganized the military, reinstating many army officers. Just in time. Hitler was also paying attention to the Russo-Finnish conflict. When he saw how the Finns had repulsed the Russians, he was convinced the German army could easily beat the USSR.

Had Stalin stayed out of Finland in 1939, the nation would most likely have remained neutral in World War II. Had Germany violated their neutrality, they would have fought as ferociously against the German invaders as they had against the Russians. Finland might have become a Russian military ally out of its own interest. As it is, Finland saw the Soviet Union as the greater threat and fought in World War II on the side of the Germans. One of the ways in which the German army used the Finnish forces was as winter combat trainers.

But Hitler failed to learn one of the main lessons of the Winter War. He would soon send his own troops into the Russian winter with the same overconfidence and lack of clothing that had been so devastating to the Soviets. The results would be very similar.

Gee, It's Cold in Russia, Part IV
Hitler Invades Russia

It is not that Hitler was unaware of Napoleon's track record when he decided to invade Russia. He simply chose to ignore it. Hitler would not tolerate any talk of Napoleon among his generals and advisers. Operation Barbarossa would not be anything like Napoleon's ill-fated invasion, he reasoned. A German victory over the USSR was all but ensured. Russia had once been a force to be reckoned with, but Hitler was well aware of Stalin's purges of generals, and he had seen the effects in Russia's war with Finland. What is more, Stalin had been reassured by the nonaggression pact between the two nations, and by the commonsense belief that Hitler was not crazy enough to start one war before he'd finished another. Blitzkrieg—the lightning-war tactics that had been so effective in other parts of Europe—would certainly topple Stalin as well. All that was needed was a single, short campaign. Russian winter was irrelevant. The Germans would not be there that long. The Soviet people, who were the victims of Communist oppression, would surely hail Hitler as a liberator. "You only have to kick in the door, and the whole rotten structure will come crashing down," Hitler said. Hitler had vastly underestimated both the will of the Soviet people and the excesses of the Russian weather.

The winter of 1941–1942 was supposed to be mild. The noted German meteorologist Franz Bauer, who was one of the first to experiment with long-range weather forecasts, assured Hitler as much. Since the preceding three winters had been especially cold, it had to be warmer. Four cold winters in a row had never been seen in one hundred and fifty years of Russian weather records.

On December 18, 1940, Hitler issued a directive that "the bulk of the Russian army stationed in Western Russia is to be destroyed in a series of daring operations spearheaded by armored thrusts. The organized withdrawal of intact units into the vastness of interior Russia is to be prevented."

Initially slated to begin around May 15, 1941, Operation Barbarossa was to be a three-pronged attack. One division would attack through the Baltic States and seize Leningrad (now St. Petersburg); a second would strike east toward Moscow; and a third would take Kiev and occupy Ukraine. The operation was delayed for six weeks when Germany was called upon to secure its flank in the Balkans, and so it began on June 22, 1941, with the single codeword "Dortmund."

It was the largest and most elaborate attack ever launched. The Soviet Union was the largest country in the world, stretching across two continents and covering nearly one-sixth of the earth's surface. It had the third largest population in the world and, most important for Hitler, tremendous natural resources. Even staying out of the "vastness of interior Russia" the Russian field of operations would cover an area several times larger than all those of Western Europe. It extended 3,200 km (2,000 mi) over mountains, forests, deserts, rivers, and swamps. It was to be a war of epic proportions.

The summer of 1941 was dry, with temperatures rising to an oppressive 40°C (104°F). Few of Russia's roads, at that time, were paved. The sun baked the earth, and the German tanks kicked up the dust. Radiators and filters became clogged. Sweltering, sunburned soldiers, dizzy from thirst, threw off extra clothing and left it behind.

Still, the first phase of Barbarossa was a great success for the Reich. The Germans, against all odds, had managed to take Russia by surprise. They quickly drove deep into the Soviet Union, crossing two-thirds of the distance to Moscow by July and capturing three million Russians. In the first two days, Russia lost more than two thousand aircraft, and within two weeks, the Red Army had lost 747,850 men. In some places, the Germans advanced fifty miles in less than a day.

The Germans had some reason to believe that the will of the Soviet people was not with their leader and that they would be hailed as liberators. Certainly, not all Soviet citizens were enamored of Stalin and Soviet-style Communism. Ukrainians, Belorussians, and the multicultural people of the Transcaucasus had their own histories, cultures, religions, and languages. The recently annexed Baltic States—Estonia, Latvia, Lithuania, and Moldova—and the Karelo-Finnish republic, wrested from Finland in the Russo-Finnish conflict, could potentially have joined the Germans in a quest for independence from Moscow. (The Finns, in fact, did.)

The Germans arrived as liberators, bearing posters declaring the end of the hated collective farm. "The free peasant on his own land! Come work with us to shorten the war!" Initially, some Soviet citizens apparently did see the German troops in this light. In one village, General Heinz Guderian's unit—the one that was slated to invade Moscow—was greeted by women bearing "wooden platters of bread and butter and eggs and, in my case, refused to let me move before I had eaten."

If you wish to keep the goodwill of the people, however, it helps if you do not treat them as members of an inferior race and culture. The German occupiers, filled with Nazi zeal, were often brutal to the "inferior" Slavs. This was a battle to rid the world of the "Jewish-Bolshevik infection." Anyone who stood in the way was to be liquidated. The Slavs may have been better than Jews, but they were still *Untermenschen* (lesser people), not members of the Aryan race.

The Germans made no provisions for their prisoners of war. They were starved until they looked "more like the skeletons of animals than humans." Within six months, more than two million Soviet POWs starved in German captivity. By some accounts, their bodies were used to fill potholes in roads. The soldiers were followed by death squads who rounded up Jews and Communists and slaughtered them.

If there had been any chance of a rebellion by anti-Communist factions or Soviet Republics against the USSR, it was irreparably quashed. Stalin looked downright fatherly by comparison. He was able to unify Soviet will and mobilize the Motherland against this brutal outside force. Russian soldiers and civilians alike rose to meet the enemy with fanatical determination. Armed civilians mounted attacks on bridges and supply depots, and in keeping with Russian military tradition, when they were forced to retreat, they poisoned wells so the Germans would suffer dehydration and illness.

On July 28 the infamous Order #227 was issued by the Soviet Supreme Command. "Not a single step back!" Russian soldiers would turn back the Germans, or die trying. Just in case they were unable to muster the courage on their own, the Red Army established penalty battalions composed of those who showed cowardice. They were sent to the most dangerous spots along the front. This was an army of men who had nothing left to lose.

In late August 1941 the German army had been so successful that Hitler ordered his generals to cease the advance on Moscow and turn toward Ukraine, where they could gobble up grain-rich lands. This delay would have serious consequences for the Germans.

Russia's secret weapon, the weather, was about to turn the tide of the war. Hitler failed to take into account one of the characteristic features of Russian weather, namely the *rasputitsa*, when the soil turns into soup. Although Hitler had learned something from the misadventures of Charles XII and Napoleon, and he expected a muddy period in spring, he believed they would already be flying the Swastika over Moscow by then, so it was of no consequence. He did not listen to the advice of meteorologists who warned him about the fall mud period.

Hitler ordered the drive on Moscow to resume on October 2, 1941, and the city went on high alert. Everyday Muscovites, including seniors and children, dug an antitank trench 98 km (61 mi) long, and antiaircraft balloons were inflated outside the Bolshoi Theatre and on Red Square. "Socially important elements" were evacuated, and Lenin's coffin was removed to the country. Meanwhile, new battalions from Siberia were moved to the outskirts of the city. Then the Germans encountered the mud.

Tanks sank up to their axles. Each step of a march became a plodding drain of energy. First dirt roads became impassable swamps, then even the gravel-top roads gave way. Supply trucks broke through the surface, leaving sinkholes. Horse-drawn vehicles, which were an important part of the war effort, were stopped in their tracks when the horses dropped dead from exhaustion, stuck hip-deep in the quagmire.

The Soviet forces also had to deal with the mud, of course, but they were better prepared. Their tanks had wider tracks than the German Panzers, and they had higher ground clearance. Plus they were fighting a defensive war, which meant they did not have to rely as much on quick movement.

The ground remained a soppy nemesis for the Germans until November brought the frost and they could start again toward Moscow. The Sixth Panzer Division got as close as 24 km (15 mi) from the Kremlin on December 1, but that night the mercury fell to –40°C (–40°F). Around this time, soldiers on the front contacted Hitler's meteorologist, Franz Bauer, and told him how cold it was outside. They asked if he was standing by his forecast that the winter would be mild. "The observations must be wrong," he said.

They were real enough to the soldiers. Under the conditions, firing pins shattered, machine guns froze, and the explosions of shells were dampened by snow. The German troops, dressed in their summer uniforms, were suffering from frostbite. The roads, once impassable because of mud, were now snowed-in.

The Russians were not immune to cold, but they were better equipped for it. The reserves now heading to the front were coming straight from Siberia. They had grown up with the cold; more important, they were dressed properly. Most Red Army soldiers had padded jackets and white camouflage suits. They wore fur caps with ear flaps and felt boots. By contrast, the German infantrymen's boots were put together with iron nails that actually sped the onset of frostbite.

Hitler was forced to approve a withdrawal. By the end of 1941, they had moved back ninety miles away from Moscow. It was now clear that Blitzkrieg would not subdue Russia and Hitler now faced a long war of attrition. To survive, he would need control of the oil fields of Baku, which would provide fuel for the Germans while keeping it from the Soviets. Along the way, Hitler's troops would seize a city strategically located on the Volga, Stalin's namesake city, Stalingrad (now Volgograd). Its capture would have maximum symbolic impact as well.

In June 1942 the Germans headed south. In the mild summer months, the Germans were once again in their element. They stormed Stalingrad with a vengeance. The initial air raids burned the predominantly wooden city to the ground and reduced the remaining buildings to rubble. A deployment of one hundred thousand Germans—twice as numerous as the Soviet contingent—then marched in.

"The Russian is finished!" Hitler announced, a bit prematurely. Once again he underestimated the ferocious tenacity of a people fighting for the survival of their nation and culture, and once again he underestimated the Russian winter.

On August 27 General Georgi Zhukov was promoted to Deputy Supreme Commander, second only to Stalin. His plan was to lull the Germans into a false sense of security, wearing them down with an "active defense" while secretly building up a huge counterattack. Timing and secrecy were key. Over the next two months Zhukov stealthily moved massive numbers of men and matériel to a position east of Stalingrad. They were aided in no small part by weather conditions that prevented accurate reconnaissance flights.

By September the Germans had reached the city's center. The Soviets, fighting in the debris of buildings, fought off as many as ten attacks in a single day. Civilians were enlisted: they fought, built shelters, and repaired damaged tanks. Soviet snipers terrorized the Germans, weakening their morale. By early November, the Germans were spent.

On November 19, under cover of a snowstorm, the Soviet troops took the offensive. They surrounded the Germans. When Hitler's generals urged him to allow the troops to break out of Stalingrad, and strike in the west, Hitler refused: "I will not leave the Volga!"

The encircled force initially numbered three hundred and thirty thousand soldiers. Of these, one hundred thousand were taken prisoner. The rest were forced to fight the elements in the rubble of the buildings they had felled. The worst of the Russian winter was ahead of them.

The Sixth Army would need 300 tons of supplies each day. The Luftwaffe was only able to deliver about 100 tons, and in the process they lost 490 planes and 1,000 crewman. As the Luftwaffe became increasingly unable to fly in supplies, the encircled Germans fell by the thousands to starvation, frostbite, and Russian snipers. The supplies that did make it to the ground were no longer distributed efficiently. Starving, freezing men scrambled for whatever they could get—it was every man for himself. As starving soldiers had done for centuries, they butchered their horses, and when that did not provide enough meat, dogs, cats, and rats became supper.

The winter freeze was a boon to the Russians. On December 16 the Volga finally froze. The Germans, short of shells, could not bombard it. Chuikov's Sixty-second Army built an innovative ice bridge across the river. Over the course of the next seven weeks eighteen thousand trucks and seventeen thousand other vehicles would cross, bringing foods, weapons, medical supplies, and warm clothes to the Russians.

Inside the ruined city, the Germans continued to suffer in temperatures that hovered around –35°C (–31°F). Surviving on only five hundred calories a day, they had no resistance to hepatitis, dysentery, and typhus. They were unable to wash because there was not enough fuel to melt the snow. Only the lice had enough to eat. Conditions were so appalling that the German military issued an order that "Suicide in field conditions is tantamount to desertion." How they intended to punish it was not clear.

"The second grim winter in that accursed country," wrote German infantryman Benno Zieser. "Completely cut off, the men in field grey just slouched on . . . from one defence position to another. The icy winds of those great white wastes which stretched forever beyond us to the east lashed a million crystals of razor-like snow into their unshaven faces, skin now loose-stretched over bone, so utter was the exhaustion, so utter the starvation . . . then the debilitated body ran down and came to a standstill. Soon a kindly shroud of snow covered the object and only the toe of a jackboot or an arm frozen to stone could remind you that what was now an elongated white hummock had quite recently been a human being."

The last German airplane flew out of Stalingrad on January 12. It contained a sack of letters from young men who knew they were doomed to death. The letters never reached their intended recipients. German propaganda minister Joseph Goebbels ordered the mail seized in case any of the contents could be used to further the cause of the war. Unsurprisingly, they found that there was little in them that would stir waves of patriotism.

"You cannot tell me that the comrades die with the words 'Germany' or 'Heil Hitler' on their lips," wrote one soldier. "There is no doubt that people are dying, but the last word is meant for their mothers or the ones they loved most, or is just a call for help."

In January 1943 the remnants of the Sixth Army sent a last, desperate request to Berlin to be allowed to break out of Stalingrad. The reply came on January 24, 1943. "Surrender is forbidden. Sixth Army will hold their positions to the last man and the last round and by their heroic endurance will make an unforgettable contribution towards the establishment of a defensive front and the salvation of the Western world."

On February 2, 1943, the Germans were finally forced to surrender and the two-hundred-day battle, fought over 100,000 sq km (3,861 sq mi), came to an end. The costs of the battle are incomprehensible. Of the three hundred and thirty thousand Germans who set out to conquer Stalingrad, ninety thousand would be taken prisoner; the rest would all fall—some to battle but most to hunger, sickness, and bitter cold. Of the prisoners, another twenty-five thousand would die in a long march through ankle-deep snow to Siberian prisons. Of those who made it to the labor camps, only twenty-five hundred would ever see Germany again.

It took the work of thirty-five hundred civilians and twelve hundred German prisoners to clear Stalingrad of the frozen corpses; bodies would continue to surface for several decades. It would be another fifteen months before the Allies, led by the United States, fully committed their forces against Germany with the D-Day invasion. But to many observers, the beginning of the end of the Third Reich happened in Stalingrad. The fact that Hitler had abandoned three hundred thousand Germans to die in an icy hell shocked the German public and the military. Germany never regained the offensive on the Eastern Front. The morale of the Germans was crushed, and the resolve of the Russians was strengthened—they would not stop until they had taken the city of Berlin.

The Russian victory came at a cost that is impossible to fathom. The Soviet Union had lost more people in the battle for Stalingrad alone than the British and Americans combined lost in the entire war. At the end of the Great Patriotic War, as the Russians call it, twenty-eight million Soviet people would be dead, including 95 percent of Russian youths between the ages of seventeen and twenty-one. More than seventy thousand villages would be burned to the ground, and at least 60 percent of the infrastructure of the nation, including forty thousand medical centers and eighty-four thousand schools, would be destroyed.

The weather was not the only factor that turned the tide of the war, but it played a major role. "Climate is a dramatic force in the Russian expanse," wrote a committee of former German generals who had served on the Eastern Front. "He who recognizes and respects this force can overcome it; he who disregards or underestimates it is threatened with failure or destruction."

D-Day

Two years after Hitler's abortive effort to invade Moscow, a Nazi sweep of Europe no longer seemed inevitable, but neither did an Allied victory. The Soviets had beaten the Germans all the way back to the Polish frontier and Allied bombers were steadily destroying German cities, but Hitler still commanded a region that extended thirteen hundred miles from the Atlantic to the Dnieper. He was also on the verge of unleashing two highly destructive weapons, the V-1 bomb and V-2 rocket. Stalin urged Churchill and Roosevelt to open a second front in Europe to take some of the pressure off the Russians. Gen. Dwight D. Eisenhower, Supreme Commander of the Allied Expeditionary Force, prepared a massive assault as the opening phase of a campaign in Western Europe. The first twenty-four hours would be critical.

Two years in the planning, Operation Overlord would be the most ambitious seaborne invasion in history. It would require nearly three million Allied troops—British, American, Canadian, Polish, French, and Czech—in four thousand ships and eleven thousand planes. Moving so many troops would require extensive planning and just the right weather conditions: the landing had to take place at dawn during low tide to reveal obstacles; there would have to be about 5 km (3 mi) of visibility for supporting naval gunfire; and the seas had to be calm. Finally, they needed a full moon to enhance the large-scale nighttime airborne operations. The weather would have to hold for at least thirty-six hours to provide enough time to land forces and secure the beachhead. Ideally, the attack would be decided on very short notice, but such a large invasion force could not idly wait around for the weather to break. A sucessful invasion of Normandy would depend on the skills of meteorologists more than any previous battle had.

From the beginning of the war, weather forecasting was seen as a vital part of military intelligence gathering. Meteorologists were employed in force for the war effort. At its peak in the summer of 1944, the Army Air Forces Weather wing employed nineteen thousand officers and other personnel in nine hundred weather stations around the world. During the early days of the U.S. involvement, a strict code of weather secrecy was imposed on radio stations by the Code of Wartime Practices. Issued in January 1942, a month after the bombing of Pearl Harbor, it instructed broadcasters to omit all mention of weather except when directed otherwise by the U.S. Weather Bureau. Although the code was voluntary, radio programmers assumed that they could lose their station licenses if they didn't "volunteer." When baseball games were rained out, sportscasters relied on euphemisms like "muddy fields" or the generic "canceled because of weather." When a major tornado outbreak hit the States on March 16, 1942, with two dozen touchdowns from Mississippi to Indiana, radio stations remained silent and one hundred forty-eight people were killed. Shortly thereafter, the censorship code was revised to allow emergency warnings—but only after the Office of Censorship had signed off on each alert.

The weather watchers for the D-Day invasion were culled from Britain's Meteorologic Office, the Royal Navy, and the U.S. Army Air Forces. They all made separate forecasts and submitted their recommendations to the chief meteorological adviser, Capt. J. M. Stagg, a tall, quiet, blue-eyed Scotsman. The Americans and Brits had different approaches to forecasting and often came up with contradictory conclusions. The Americans relied on past records to see if the weather fit a past pattern. The British used physics and mathematical models—but without computers, it was difficult to perform the complex calculations for a twenty-four-hour prediction.

"The outcome of D-Day, perhaps the whole future of the Western world rested on these forecasts, so I think you could say there was some pressure," Dr. Lawrence Hogben, one of the Royal Navy forecasters, told the *Daily Telegraph* in 2004.

It was agreed that the ideal time for an invasion would be the spring. Often in the spring, air masses from the Azores drift toward northeast forming a "blocking high" over Normandy which, along with a split in the upper-air jet stream, guides surface storms off toward Scandinavia. That means there is fair weather about 40 percent of the time in April and about 30 percent in May. Then, in June, decreasing pressure of the Siberian high (which permits a greater number of cyclones to track through the area) tends to bring storms and rough seas. So the operation was initially planned for May, but planners felt that additional air operations were needed and the size of the invasion force should be increased. Operation Overlord was postponed until June.

Three dates were selected. It would take place either June 4, 5, or 6—ideally on June 5. If weather did not permit the invasion to take place then, the next window of opportunity would be on June 19.

As June 4 approached, so did a storm. General Eisenhower

postponed the operation for the next day. As the day progressed it became clear that the sky was not, and the meteorologists were not confident that there was going to be any change. Then a Royal Navy ship off the coast of Iceland reported sustained rising pressure. This could mean that a ridge of high pressure was developing behind the cold front that was over the Channel. It might provide a window of favorable conditions during the important dawn hours of June 6, but the weather would likely be marginal. The three forecast centers were split on whether or not to proceed. They finally agreed to recommend the attempt, by a vote of 2–1.

With all the best forecasters on the job, the weather is still prone to surprises. On June 5, 1944, Capt. Harry C. Butcher, a naval aide to Eisenhower, recorded some of the behind-the-scenes discussions of the "weather dope."

"Monday morning the weatherman who had spoken for all the weather services, after a rather doleful report, was asked, 'What will the weather be on D-Day in the Channel and over the French coast?' He hesitated . . . for two dramatic minutes and finally said, conscientiously and soberly, 'To answer that question would make me a guesser, not a meteorologist.'"

After much reflection and debate Eisenhower finally said, "Okay, we'll go." On the evening of June 5, the Allies knew they were either on the cusp of a victory that would turn the tide of the war or about to encounter crushing defeat. "I am very uneasy about the whole operation," said Sir Alan Brooke, chief of the Imperial General Staff. "It may well be the most ghastly disaster of the whole war."

Eisenhower even wrote a message to be published in the event the mission went astray. "Our landings in the Cherbourg-Havre have failed to gain a satisfactory foothold and I have withdrawn the troops. My decision to attack at this time and place was based upon the best information available. The troops, the air, and the Navy did all that bravery and devotion to duty could do. If any blame or fault attaches to the attempt it is mine alone."

Fortunately, the timing turned out to be ideal. The marginal weather gave the Allies an important element of surprise. The stormy conditions on the fourth and fifth had prevented German naval patrols and reconnaissance planes from surveying the invasion preparations. But the Germans were confident that the same weather that kept their planes in the hangars would prevent an Allied attack. The Allies had also conducted a very effective disinformation campaign, leading Hitler to believe the invasion would come farther up the coast at Pas-de-Calais. The Allies had created this impression by repeatedly bombing the Calais area as if to ready it for an invasion. They also sent out messages about the movements of fictional divisions. The Americans had the "First United States Army Group," and the British had the "Fourth Army," a phantom regiment that was supposedly in Edinburgh planning to invade Norway. They even went so far as to create a dock, surrounded by inflatable tanks, fake warehouses, and empty barracks to fool aerial surveillance. Field Marshal Erwin Rommel, the commander of Germany's troops in northwest France, was so confident that the Allies would not invade in that place or time that he took a quick trip home for his wife's birthday on June 6.

The invasion plan called for five divisions to land along an 80 km (50 mi) front. British and Canadian forces would land to the east, American to the west. At 6:30 A.M., five thousand vessels appeared in the English Channel. A sailor aboard the Canadian minesweeper *Canso* described it this way: "As far as the eye could see, there were ships. I always said that if you could jump a hundred yards at a clip you could get back to England without even wetting your feet."

As predicted, the weather had improved, but it was far from

perfect. Cloud cover the night before had thwarted some of the planned air assaults. The wind churned up waves as great as five feet high. They crashed over the sides of the landing ships, leaving soldiers seasick, cold, and loaded down with heavy, saturated battle gear. The landing of the Third Canadian Division at Juno Beach was delayed an hour and a half by choppy seas. By the time they arrived, the element of surprise had been lost. Still, the combined British and Canadian forces took their three beaches and advanced about three miles inland toward Caen at a cost of an estimated 3,000 British and 946 Canadian lives.

The losses were greater still on Omaha Beach, where German forces on high ground fired down on the invading Americans. Landing craft were kept from the beach by barbed wire and other obstacles and often soldiers had to jump into neck-deep water. Many amphibious craft were forced to turn back. The armored units were especially hard hit. Sherman tanks, outfitted with devices that were supposed to keep them afloat, foundered in the choppy waves. Of the first thirty-two launched, twenty-seven sank to the bottom along with their crews.

American soldiers' bodies cluttered the beach. To Gen. Omar Bradley, the carnage seemed to be "an irreversible catastrophe." But by the afternoon he got the message that the troops had advanced to the ridge of cliffs that overlooked the beach and were moving inland. Those cliffs were captured at a great cost. Among the Americans, 1,465 were killed, 3,184 were wounded, and 1,928 were missing; the Germans lost an estimated 4,000 to 9,000 men.

"Not much would have to have changed for D-Day to have been a failure," Dr. Hogben said, "and a failure caused by the weather."

Had the Allies not used the window of opportunity on June 6, they might well have lost the element of surprise, even with the inflatable tanks. As it happens, on the backup dates of June 18 and

19, the coast of Normandy was pounded by a four-day storm that prevented the landing of supplies along the beach. Had the Allies been able to land, they would have had no possibility of resupply and reinforcements. Thus they would most likely have had to delay even longer. The German aerial surveillance would surely have discovered that Normandy was going to be the site of the Allied invasion, and Hitler might have used the extra time to further develop his V weapons program.

Instead, the Allies met their goal. They secured the beaches, giving them the entrée that would allow them to sweep across France. By the end of July more than eight hundred thousand Americans had landed in France. More than eighty thousand trucks bearing food, ammunition, and supplies had also come ashore, and Germany would now have to fight the land war on three fronts: the Soviet Union, Italy, and France.

Blooming with Atoms

A New Cloud Formation, the Mushroom

The terrible irony of the cold war is that, while the two superpowers never unleashed their nuclear arsenals on each other, they were both highly effective in irradiating their own people in the quest for atomic dominance. Only two nuclear bombs have ever been used in war, but 1,745 nuclear tests have been conducted by the United States and the USSR. since the first bomb was dropped. Other nuclear powers (France, Britain, and China) have detonated another 288 nuclear devices.

Data from a federal study conducted in 1998 by the U.S. Department of Health and Human Services estimates that nuclear weapons tests across the globe probably caused about fifteen thousand cancer deaths among Americans born after 1951 and another twenty thousand nonfatal cancers may be attributable to the tests. Studies now reveal that every person in the United States was exposed to increased radiation. And that's just the Americans. A Danish scientist, Asker Aarkrog, estimated in a research paper presented at an international symposium in Vienna in 1995 that nuclear activities in the Soviet Union may have increased the annual radiation dose received by the world population by one-sixth. Blame it all on the wind, which has an inconvenient habit of not remaining in one place.

It all began on July 16, 1945, at the Trinity test site in New Mexico. After twenty-eight months of planning at the top-secret Los Alamos research facility two hundred miles to the north, the first atomic bomb "was exploded in a weather situation as foul as any meteorologist could imagine for such a monumental event. Rain fell in torrents, accompanied by gusty winds, lightning, and thunder," as Benjamin Holzman wrote in *Weatherwise* magazine.

Eventually the storm passed, and the bomb was successfully exploded, giving the few Manhattan Project scientists on hand a view of a new weather phenomenon: the first nuclear mushroom cloud. When a nuclear explosion occurs, it produces intense heat. Any material close to the point of the explosion is vaporized and forms a gas. As the fireball expands and cools, it rises like a hot air balloon, taking with it the vaporized material and any dust and ash that is light enough to be drawn into the rising cloud. These are mixed with the radioactive by-products of the bomb and become radioactive themselves. As the cooling fireball rises to become a cloud it trails a "stem" of super-heated dust and debris—the now-familiar mushroom shape. And soon the bombs dropped on Hiroshima and Nagasaki put the world on notice that we had entered the nuclear age.

In 1946 America moved its proving ground to Bikini, an atoll in the Pacific. After a while, the costs of such a distant test site became problematic. So a committee was assigned to choose a location for the first aboveground nuclear weapons tests in the continental United States. Those with an understanding of the weather, notably the Army Air Forces staff meteorologist, suggested an East Coast site would be best because "the United States is predominantly under the influence of westerly winds." The wind experts, however, were outvoted.

In August 1949 the Soviets announced their nuclear status to the world with their own test. Led by Igor V. Kurchatov, "the Russian Oppenheimer," Moscow's equivalent of the Manhattan Project culminated in a 22-kiloton bomb dubbed RDS-1 for *reaktivnyi dvigated Stalina* or "Stalin's rocket engine." In the Western atomic establishment it was called "Joe 1" (also for Joseph Stalin). Stalin is reported to have wanted the Soviet test to be as terrifying as possible. The aboveground explosion destroyed homes for three miles around the Kazakhstan test site.

Yuli Khariton and Yuri Smirnov wrote in the *Bulletin of the Atomic Scientists,* "At that dramatic point, when the threat of atomic attack hung over the Soviet Union and millions of human lives were at stake . . . it was first necessary to perform a truly heroic feat that required nationwide mobilization. . . . All this was done in a country devastated by the war."

Soviet scientists had strong motivations not only to catch up with the Americans in the nuclear race but to overtake them. In a nation that had suffered as much loss as it had in World War II, the consequences of being unprepared for attack were clear and fresh in every Soviet mind. Further, the possibility of reward or punishment under Stalin could not compare to that in America. There is a story that Lavrenti Beria, the head of the Soviet atomic program, decided what rewards to bestow on successful scientists based on how severe the consequences would have been had they failed. Those who would have been shot were awarded the title "Hero of Socialist Labor," those who would have been sent to a Siberian prison were given the "Order of Lenin" and so on down the line. This story is no doubt apocryphal, but it does serve to illustrate how important it was to present Stalin with favorable results.

As to the Americans' motivation—Stalin with nukes, need I say more? The central belief when it came to testing was that "fallout is much less dangerous than falling behind the Russians." Thus began the escalation of missile making and testing. The Los

Alamos scientists were churning out an average of five new warhead designs a year. In keeping with this policy, the military strategists rejected the notion of an East Coast test site and, in order to reduce travel time and cost, chose one close to the New Mexico research facility. The obvious location was the vast open desert outside Las Vegas, Nevada. As Gerard DeGroot wrote in *History Today* magazine, "There are few other places in the United States where a fifty-kiloton bomb has little noticeable effect on the landscape. Nevada is proof that man's bomb is big, but God's earth is bigger."

What is more, Nevada's desert climate had another interesting side effect. Many Nevadans actually wanted nukes in their backyard. Agriculture and industry did not often spring up in the desert—and brought little income for the state—but nuclear tests would. Sen. Pat McCarran of Nevada actively campaigned to get the test site, and in fact, the atom was just the thing to energize the area economy. By the mid-1980s about twenty thousand people in southern Nevada were employed directly or indirectly by the testing program, which made the test site the second-largest employer in the state. Bomb tests were also tourist attractions. Hotel operators in Las Vegas, where the flashes were visible, organized package tours. The Sands casino sponsored a Miss Atomic Bomb pageant, the Flamingo Hotel salon offered a mushroom cloud-shaped hairdo, and casino guests could drink an "atomic cocktail"—a mix of vodka, brandy, champagne, and sherry. Prostitution is legal in Nevada, and the program also brought work to prostitutes who earned a living in brothels bordering the test site.

After the first detonation in the Nevada desert on January 27, 1951, Gov. Charles Russell said, "It's exciting to think that the submarginal land of the proving ground is furthering science and helping national defense. We had long ago written off that terrain as wasteland, and today it's blooming with atoms."

Whereas the U.S. tests were a boon to Nevada tourism, the Soviet tests, in Kazakhstan, were a closely guarded secret, even to those closest to the blasts. Some of the soldiers working close to the Soviet test sites learned of radiation dangers only through the revelations of Western intelligence. During the period of Soviet nuclear testing, Kazakhstan was riddled with some five hundred nuclear explosions. Soviet scientists studied the health effects of radiation on six thousand residents. While they made regular notes of their conditions, they did not treat them.

The main health concern in the United States was not among the immediate test site neighbors—there weren't many—but among those who never even suspected they were at risk, people far from ground zero who were visited days later by radioactive clouds and wind. The people most directly affected turned out to be the residents of eastern Idaho, five hundred and fifty miles north of the Nevada proving grounds.

When an atomic cloud acquires the density of the surrounding air, it stops rising and is subject to the same wind patterns and rain activity as any other cloud. The normal circulation of weather systems determine exactly where those radioactive particles will eventually come to rest. With winds blowing from south to north, Idaho was the ultimate destination of most of the high-altitude winds and rain from the Nevada blasts.

Yet weather patterns are never quite that simple. "Hot spots" erupted in many unpredictable areas as passing clouds gathered radiation, moved along, then dropped the fallout as rain several states away. Portions of the Midwest and even New England received high doses of iodine. One hot spot was in Rochester, New York. After a 1951 snowstorm, employees at a Kodak plant tested the snow with Geiger counters and found readings of radioactivity twenty-five times above normal. In 1953, after another test, there was a rainstorm in Albany, New York. Students at a local college took Geiger counter readings of puddles and found they were one thousand times above normal. The Atomic Energy Commission evaluated the Albany rain and recommended that since hot spots were more likely to occur in months with high precipitation "total fallout in the United States could be reduced somewhat by scheduling test series in the fall." The recommendation does not seem to have been followed.

British scientists believe that fallout from Soviet tests was carried in the jet stream and created hot spots in England. They suspect that some cancer deaths in the United Kingdom could be attributable to the Kazakhstan tests. The extent of radioactive contamination from Soviet tests remains largely unknown, but by all accounts Kazakhstan has abnormally high rates of cancer, cerebral palsy, and birth defects.

Scientists are divided about the health impact of this exposure. The National Cancer Institute forecast that the fallout could cause anywhere from ten thousand to seventy-five thousand extra thyroid cancer cases nationwide, mostly among people now in their forties and fifties. Other scientists, however, say the cancer and radiation connection is inconclusive.

As to the Nevada test site, it is going to be harnessed for power—wind power. About half of the former bombing range is slated to become a wind-power facility, with 325 wind turbines generating 260 MW of energy.

Sunshine over Hiroshima

"Cloud cover less than three-tenths. Advice: bomb primary."

It was fine summer weather, on August 6, 1945, in Hiroshima, a city on the southwestern coast of the main Japanese island of Honshu. It was hot and humid but clear and sunny. The city had, so far, been spared the firebombings that had destroyed many Japanese cities because several tributaries of the Ota River cut through the city hindering fire's spread. That morning, eighty-nine hundred local schoolchildren were out helping to clean and widen the streets as part of the war effort.

At 7:09 the air-raid sirens went off. The alert was triggered by a single plane which was spotted flying high above the city. It flew over and vanished without dropping a bomb. At about 7:45 A.M. the all clear sounded, and the people of Hiroshima left their shelters and went back to their daily tasks. What they did not know was that the small aircraft that had passed so uneventfully had actually sealed their fate and brought about a decision that would make the name of their city forever synonymous with nuclear war.

Gen. Leslie R. Groves, director of the Manhattan Project, had wanted Kyoto, the ancient capital of Japan, to be the target of the first nuclear bomb used in war. Leveling Kyoto would have a psychological impact that none of the other possible sites could provide. Grove was overruled, however, by Secretary of War Henry Stimson, who felt that that centuries' worth of cultural and religious artifacts in Kyoto should be preserved.

On July 30 Stimson cabled President Harry Truman, who was at the Potsdam Conference, asking for the order to drop the atomic bomb. Truman wrote back in longhand, "Suggestion approved. Release when ready."

An approaching typhoon on August 1 briefly delayed the attack, but the skies cleared in a few days. By August 5 the fliers were ready to go, and the bomb, named "Little Boy," was loaded into the bomb bay of a B-29, which its commander had named for his mother, Enola Gay.

Before Army Air Forces Col. Paul W. Tibbets flew into history in the cockpit of the *Enola Gay,* the target list had been whittled down to four: Hiroshima, Kokura, Nigata, and Nagasaki. Hiroshima, an industrial city with a population of about three hundred and fifty thousand, had been chosen as the primary target. It was surrounded by mountains, which would focus the blast's force for maximum effect. But the final decision would be made by the weather. If it was cloudy over Hiroshima, then the site with the clearest skies would be the new primary target.

The *Enola Gay* was not the only aircraft to take part in the bombing mission: three weather planes were dispatched ahead of the bomber. The aircraft that set off the air-raid sirens in Hiroshima was the *Straight Flush,* commanded by Maj. Claude Eatherly. Eatherly saw that the weather was clear over the target, radioed back, and returned home. The *Enola Gay*'s navigator, Capt. Theodore "Dutch" Van Kirk, laid in a course for Hiroshima. As the bomber approached the city, no more air-raid sirens went off. Perhaps the civil defense monitors had been lulled into a false sense of security by the earlier airplane. What was about to happen came without warning.

As the bomb was released, the aircraft, now relieved of its weight, jumped. Tibbets went into a sharp turn, a move he had practiced many times. The escape maneuver would protect the aircraft from the bomb's blast. Forty-three seconds later, the bomb detonated 1,890 feet above the ground.

There was a blinding flash of white light. A blast that was the equivalent of twenty thousand tons of TNT released a fireball that was hot enough to melt steel and granite. The shockwave leveled everything within three miles. Then the mushroom cloud appeared and rose toward the sky, lifting a swirling wake of dust and debris. Whirlwinds tore through the city and black rain fell. Glass and mirrors for miles around were shattered by shockwaves, and shards were scattered into anything in their path.

The death count is imprecise because the town's records were vaporized in the blast, but an estimated one hundred thousand people died instantly from the firestorm. Those who were not killed were burned beyond recognition. Some had lost all facial features and walked like zombies, their arms outstretched to keep the burned skin from scraping against their own bodies. Some fumbled blindly with eyeballs melted in their sockets. Because light colors reflect heat and dark ones absorb it, some of the victims had images of their clothing seared onto their flesh. Another one hundred and forty thousand were condemned to slow death from radiation sickness.

There was little help for the sick and wounded. The hospital was left standing, but more than half of the doctors and nearly all the nurses had been killed. Those who remained were seriously wounded themselves. Only a handful of physicians was left to do what they could to help the victims and catalog the effects of a new sickness.

Above it all was the crew of the *Enola Gay*. The copilot, Capt. Robert Lewis, recounted his experience in the aircraft's mission log: "The flash was terrific. . . . [T]here was without doubt the greatest explosion man has ever witnessed. I am certain the entire crew felt this experience was more than anyone human had ever thought possible. It just seems impossible to comprehend. Just how many did we kill? I honestly have the feeling of groping for words to explain this or I might say 'My God, what have we done?' If I live a hundred years I'll never quite get those minutes out of my mind."

Francis Marbert, the flight engineer of a B-29 Superfortress whose mission was to draw enemy fire away from the *Enola Gay* that morning, recalls that even twenty miles away from the target, he could see and feel the explosion. According to Capt. Lewis's account, the mushroom cloud remained visible "even after an hour and a half, four hundred miles from the target."

The United States waited for word of a Japanese surrender. Yet as Gen. George C. Marshall would later note: "What we did not take into account was that the destruction would be so complete that it would be an appreciable time before the actual facts of the case would get to Tokyo."

Perhaps there would have been time for Tokyo to receive and digest the news of the fate of Hiroshima had it not been for another meteorological event. When the Japanese failed to surrender immediately, the United States began preparations for a second atomic demonstration. The bombing raid was planned to take place on August 11, but meteorological reports indicated bad weather would hit Japan at that time. The operation was moved forward to August 9.

On August 8, the second nuclear bomb, known as "Fat Man," was placed into the bay of a B-29 called *Bock's Car*. At 3:47 A.M. pilot Maj. Charles W. Sweeney took off headed toward Kokura, a city that contained a major weapons arsenal and Nippon Steel's main works. Sweeney had not yet dropped a bomb on an enemy target during the war, and in fact, he would only ever drop one. He had, however, taken part in the Hiroshima mission. When the first atomic bomb was released from the *Enola Gay,* Sweeney's craft, known as *The Great Artiste,* followed and dropped a series of sensors designed to collect scientific data.

Kokura had already been attacked by conventional weapons the day before. The residents could not have guessed that this attack, which killed two thousand, actually saved their city from an even greater nightmare.

A weather reconnaissance aircraft that flew ahead of *Bock's Car* reported that the target could be seen. By the time the bomber arrived, however, the smoke was still rising from the fires of the previous day's attack and combined with ample cloud cover to obscure it. The overcast skies prevented Sweeney from seeing the target. *Bock's Car* circled the city three times with its bomb doors open. Their fuel started to run low, so Sweeney decided to turn toward the backup target, Nagasaki.

Sweeney's fuel reserves were so critical at this point that he would have only one chance to drop his payload. It seemed as though the mission might have to be aborted, as Nagasaki was also cloudy that day. Then, at 10:58, the bombardier spotted the target through a break in the clouds and released the bomb.

William L. Laurence of the *New York Times* witnessed the rising mushroom cloud. The observers on his aircraft "saw a giant ball of fire rise as though from the bowels of the earth, belching forth enormous white smoke rings. . . . [W]e watched it shoot upward like a meteor coming from the earth instead of from outer space."

"Fat Man" was a more powerful bomb than "Little Boy," and its effect was even more devastating to the affected part of the city. Hills flanking the Urakami River shielded the harbor and historic districts from the blasts, but those same hills focused the blast, creating devastation in the Urakami Valley. An estimated twelve thousand buildings were destroyed by the explosion and the fires that followed in its wake. "Fat Man" unleashed incredible heat, about 7,200°F. The flash of light generated by the explosion cast shadows, and the intense heat etched those shadows into place on walls and buildings. Charred bodies floated in the river.

One Nagasaki survivor whose account was to be part of a Smithsonian Institution exhibit that was later published in the book *Judgment at the Smithsonian*, recalled, "[B]odies were piled so high [that] surely another could not have been added. Neither dead nor living, nor male and female, could be distinguished among the overlapping bodies. . . . Their hair was burned crisp and wrinkled; their clothes were in tatters . . . and some kind of pitch-black substance, like coal tar, stuck to their heads and bodies."

Another survivor recalled seeing a boy, frozen in a running position, like a statue. On a tree beside him was a dead cat. "[I]ts body was covered in the scorched and frizzled remains of fur. Without disintegrating or falling from the tree, it glared with eternally locked eyes in the direction of the boy."

The official death toll in Nagasaki was about seventy thousand people. Each year, the citizens of Kokura who were in the city on that fateful day meet to celebrate the clouds that spared so many of their lives.

Misreading the Monsoons

The battle of Dien Bien Phu has been called "a little Stalingrad." It was warmer in Dien Bien Phu, but as in the World War II siege, an invading army found itself encircled and left to battle nature as supplies ran low and the enemy got closer and closer.

At the end of World War II, the European powers jockeyed for position in the new world order. For France, this meant reasserting power in its neglected Asian colonies. The surrender of the Japanese had created a new balance—or imbalance—of power in the region.

Vietnamese Emperor Bao Dai had declared independence from France with the backing of the Japanese, but the world knew that the Japanese were in no position to stand behind their support. With France still war-weary, the nationalist leader Ho Chi Minh decided to seize the moment. Ho knew quite a bit about his soon-to-be enemy, the French. He had left Vietnam in 1911, working aboard a French liner. From there he traveled to London and the United States and settled for a time in France, where he had the time to become a founding member of the French Communist Party in 1920. He went on to study revolutionary tactics in Moscow before returning to his native land to found the Communist Party of Indochina. During the war, Ho's independence movement, the Viet Minh, raised an army to fight the Japanese. The guerrilla army's general, Vo Nguyen Giap, admired Napoleon and had studied his tactics.

In September 1945 Ho announced the creation of the Democratic Republic of Vietnam. At first, Ho agreed that it would be an autonomous state within the French Union, but the peace did not last long. In November 1953 Gen. Henri Navarre and his forces parachuted into Dien Bien Phu, located in Vietnam's largest valley. It lies in a patch of jungle and rain forest between low mountains. Only one decent road ran through the area; therefore, the French figured it was the perfect place for a base of operations, since any large force moving to or from Laos would have to pass through eventually. They certainly never expected the Viet Minh would avoid the road and drag their artillery over the mountain by hand.

The French had been in Indochina long enough to know all about the monsoons—or so they thought. The word *monsoon* comes from the Arabic word *mausim,* which means "season." The "seasons" that affect East Asia are alternating winds that come from the south and southwest in spring and summer and the north or northwest in the fall and winter. The first set of wind brings a rainy season; the second a dry season.

The monsoons were not a problem in the November dry season and although the French were well aware that the wet season was coming along with more than 150 cm (60 in) of rain, they were not concerned. They followed the same logic that had worked so well for Napoleon and Hitler in Moscow: the weather would not be a factor because the war would be won before the seasons changed.

Navarre established a base that included fourteen thousand soldiers and, according to Richard Cavendish of *History Today*, two mobile brothels. "The French built nine outlying strongpoints in the valley," he wrote, "and gave them women's names, allegedly after their commanding officer's mistresses."

It was quickly apparent, however, that they had underestimated their opponent. Giap had a force of fifty thousand men armed with Chinese weapons, including American howitzers captured in Korea. They surrounded Dien Bien Phu, which meant the only way the French would be able to restock their supplies was by air.

At the same time, Ho Chi Minh's army was using the jungle brush for cover and moving soldiers and heavy equipment onto the forward slopes of the hills of Dien Bien Phu. The vegetation foiled French intelligence as aerial reconnaissance revealed nothing but greenery, even though the slopes around the French outpost were loaded with enemy forces waiting to strike.

General Navarre was getting worried. The spring monsoon season was quickly approaching, and he knew the rains would bring flooding. The Nam Youm River would overflow and cut the camp in two—visibility would be poor, airplanes would be grounded, and communication might be cut off. If that happened, the soldiers would be completely isolated without supplies and with little ability to defend themselves. French command in Hanoi listened to his concerns, but they were of the opinion that the monsoons would thwart the Viet Minh rather than their own forces.

The French public, for the most part, had other things to worry about. They were still rebuilding their cities and their economy after the war against the Nazis, and the small conflict in a faraway colony did not capture the French imagination. In May 1953 a poll of newspaper readers revealed that only 30 percent ever followed news about Indochina. Another poll, in February 1954, found that only 8 percent of respondents supported French intervention in Indochina. "[T]he Indochina War remained for its duration, in large part, 'the forgotten war,'" wrote historian David Drake.

On March 13, 1954, the Viet Minh began their siege. The monsoon started early that year, and by late March heavy rains were flooding French trenches. The sandbags in the fortifications of Dien Bien Phu were saturated and heavy. Many of the bunkers burst into torrents and floated away as timbers in muddy water. Stagnant pools brought disease to the soldiers.

Just as Navarre had predicted, the Nam Youm River rose up and flooded parts of the base, filling many of the dugouts to their brims. "The situation of the wounded is particularly tragic," Navarre wrote. "They are piled on top of each other in holes that are completely filled with mud and devoid of hygiene."

The thick cloud cover forced the French aircraft to fly low, where they made easy targets for the Viet Minh antiaircraft guns. Yet the monsoon did not seem to be interfering much with the Vietnamese. It just gave them more cover as the lush vegetation grew thicker. Each plane they brought down or chased away meant less medicine and food for the French troops. The runways became too dangerous for landings, and the aircraft had to drop food and weapons. Sometimes they missed and inadvertently supplied the enemy. After the Viet Minh took the strongpoints called Gabrielle and Béatrice in March, the French artillery chief, Col. Charles Piroth, was so distraught that he committed suicide with his own grenade. By the end of April, the French defense area had been reduced to 5 sq km (2 sq mi). Nearly two thousand French soldiers died at Dien Bien Phu, and another six thousand were wounded or ill. Casualties on the Vietnamese side are estimated at ten thousand.

On May 7, 1954, the last seven thousand men—hungry, exhausted, and rain-soaked—surrendered to the Viet Minh. Of those, only about three thousand survived the conditions in the prison camps. It was the end of French power in Indochina, and it sent shockwaves through colonized nations around the world.

Dewey Defeats Truman

It is one of the most famous photographs in American political history. The president-elect, Harry Truman, grins triumphantly as he holds up a copy of the *Chicago Tribune* with the blazing headline: DEWEY DEFEATS TRUMAN. Collectors of historic newspapers pay as much as nine hundred dollars for this gem of failed forecasting.

Thomas Dewey, of course, did not defeat Harry Truman for the presidency. The *Chicago Tribune* made the mistake of believing polling data, which said there was little chance Truman could win. The day before the election, Gallup predicted Dewey would get 49.5 percent of the vote and Truman 44.5 percent. The Crossley poll came up with almost the same result. The Elmo Roper poll showed an even more impressive 52.2 percent for Dewey and 37.1 percent for Truman.

How did the pollsters get it so wrong? There were, of course, many factors, which historians have discussed and debated for decades. One of these was the weather in Illinois.

Weather and politics have a long history, going back to the days when political office was won on the tip of a sword rather than at the ballot box. The word *campaign* itself derives from a military term that can be traced back to the days when armies stood down during the freezing cold and ventured out into "the field" only when weather permitted. The Latin word for "open field" is *campania* and this passed into fourteenth-century English as *champaign*. (The French wine was named for a region of open fields that derives its name from the same Latin root.) *Champaign* evolved into *campaign*. The word for the field was metaphorically used for the military taking of the field, and eventually this sense was expanded to include any attempt to mobilize a large number of people, hence a political campaign.

When people began to select their leaders through elections, the effect of the weather became less direct. Yet inclement weather continued to play its role by hindering campaign efforts and keeping the less-motivated people from the polls. When the race is very close, as it was in the case of Dewey vs. Truman in 1948, a few people's deciding not to venture into the elements is enough to swing the election.

In 1948 splits within the Democratic party made a Truman victory seem like a long shot. Two spin-off parties were created, one on the left, and one on the right. On the left was the Progressive Party, led by Henry Wallace. Wallace and his followers held Truman responsible for the cold war with Russia, while their candidate favored negotiations to lessen the tensions. The Progressive slogan was "One, two, three, four, we don't want another war."

On the right was South Carolina governor J. Strom Thurmond and his States' Rights Democratic Party. This party was created after Truman proposed a series of measures that would guarantee equal rights to African Americans. The proposals were controversial among the so-called Dixiecrats, or Southern Democrats. When Truman announced his plan, thirty-five delegates walked out of the Democratic Convention to form their own party. The States' Rights delegation did not expect to win, but they hoped to get enough votes to throw the election to the House of Representatives, where they could swing the ballot to a candidate who opposed civil rights legislation.

All this Democratic infighting was a boon to Republican Thomas Dewey. The *New York Post* wrote: "The party might as well immediately concede the election to Dewey and save the wear and tear of campaigning." Dewey was so confident in his assured win that he mounted a very subdued campaign. Harry Truman, on the other hand, embarked on a thirty-one-thousand-mile "whistle stop" train trip across the nation and delivered hundreds of speeches.

Democrats have often said that the rain favors them because it presents more of a challenge to rural (conservative) voters than to the urban (liberal) voters. On election day 1948 this was borne out—a storm system in the lower Mississippi Valley spread rain across Illinois with especially heavy downpours in the South. Not only was the rain more likely to muck up dirt roads than urban transport, the storm itself was concentrated over the rural section of the state, while it mostly spared Chicago and the industrial north. Meanwhile, a Pacific storm brought heavy rain to California. It fell in the predominantly Republican northern part of the state, while the Democratic southern part had sunshine.

A difference of 29,294 votes—or .28 of one percent of the electorate—in Illinois, California, and Ohio could have changed the outcome. The rain in Illinois and California may well have tipped the scales. In the end Harry Truman captured 303 electoral votes; Thomas Dewey, 189; and Strom Thurmond, 39.

Canadian Chill Saves a National Park from Nuclear Contamination

Wapusk National Park, which takes its name from the Cree word for *white bear*, is one of the world's largest known polar bear breeding grounds. On average, 190 female bears remain on the shores of Cape Churchill each year to have their cubs. A treeless tundra of willows, cotton grass, and tamarack moss, Canada's seventh-largest national park is part of a large wildlife area set aside to maintain the ecological integrity of the Hudson James Lowlands. It is also an important destination for whale watchers, botanists, and geologists. Archaeological research of the area has uncovered important information about the indigenous North Americans and their nomadic culture. Home to the Dene and Cree tribes, the region was once a center of the fur trade. And were it not for the harsh, subarctic climate, Wapusk would have been the test site of Britain's first operational nuclear bomb, the 25-kiloton Blue Danube.

Still reeling from World War II, with a shattered economy and the threat of a new war in Europe against the Soviet Union, the United Kingdom felt it would not be prudent to leave its nuclear deterrence in the hands of its ally the United States, "special relationship" notwithstanding. Following the example of its nuclear predecessor, the British came to believe that only the threat of massive retaliation and mutually assured destruction could create security for the nation. England had many capable scientists, many of whom had made valuable contributions to the Manhattan Project. Testing, however, was going to be a problem. Unlike the United States and Soviet Union, the United Kingdom did not have vast, lightly populated areas. There was no way a nuclear device could be safely tested on the British Isles.

So the British looked elsewhere. A twenty-page document, "Technical Feasibility of Establishing an Atomic-Weapons Proving Ground in the Churchill Area," was declassified in 1994, revealing the plan for detonations in Canada. The Churchill plan was penned by C. P. McNamara, of Canada's Defence Research Board, and William George Penney, an official in Britain's Ministry of Supply, known as the Oppenheimer of Britain (he could also have been called the Igor V. Kurchatov of Britain). Had the plan become operational, as many as twelve atomic devices would have been detonated at or above ground level at a site near the mouth of the Broad River inside what is now the Wapusk National Park. There was also discussion of making the area available for U.S. tests.

Canada was then a significant military power—after World War II, it had the world's fourth-largest navy and well-equipped ground and air forces. Only eight kilometers east of the proposed proving ground, Fort Churchill housed six thousand Canadian and U.S. soldiers. The report's authors felt that this region of Canada was "valueless . . . a wasteland suitable only for hunting and trapping." It was ideal, they said, because prevailing winds came from the north and the site was south of the inhabited area.

In the end, however, the British decided that Churchill,

Manitoba, was simply too cold and uncomfortable. The ice and snow would make it difficult to maintain a reliable landing strip. Much more appealing was the warm, hospitable climate of Australia's Monte Bello Islands.

Of course there were other factors in the decision as well. It is now known that there were Soviet spies operating near the Churchill military base. A British nuclear test in Canada would not have been a secret to the Kremlin, but Australia was distant enough to be off the metaphorical Soviet radar. Australian Prime Minister Robert Menzies also made it very easy for the British. According to Judge James McClelland, who presided over the Australian Royal Commission of Inquiry into the effects of the tests, Menzies "just said yes" and did not even consult his cabinet before giving approval.

Had Canada been the proving ground, the tests would not only have devastated the wildlife refuge, but would likely have blown fallout southeast toward Toronto, Montreal, New York, and maybe as far as Scandinavia.

Instead, Australia was the site of Britain's first nuclear test device, "Hurricane," which was exploded on October 3, 1952, inside the hull of a British naval frigate off Trimouille Island. By the time the British finished their tests in Australia in 1958, a dozen nuclear devices had been detonated and hundreds of "minor trials" involving radioactive materials had been conducted.

The test that is believed to have had the greatest impact on the Australian population was Operation Buffalo, the first test series carried out on the Great Victoria Desert in South Australia. Prior to the test, a one-page item published in the *Adelaide Advertiser* on May 16, 1956, quotes a professor E. W. Titterton: "An Australian Safety Committee comprised of six University and Defence scientists has the hazard problem under continuous review and is responsible for choosing firing times so that the weather conditions are favorable and no damage to life and property can result either on the mainland, to ships at sea, or to aircraft."

The same newspaper, twenty-four years later, reported that Adelaide City, in fact, experienced widespread fallout when a secondary plume of radiation from the tests drifted south and rained out over the city's 518,000 inhabitants. A 1980 report says that Adelaide had measured radiation levels of nine hundred times above normal, although the same report states that "the levels were very low in human effect terms."

Not everyone agrees with this statement. British and Australian servicemen who claim their health was impacted by the tests have been trying for years to get compensation from the British government. A 1984 report found that the Maralinga test had posed a serious hazard to the Aboriginal Maralinga Tjarutja community, and they received about nine million dollars to compensate them for the loss of their land.

The most high-profile potential fallout victim may have been British Prime Minister Tony Blair. When he was three years old, young Tony lived with his family in Adelaide, about 350 miles south of a test site. An unanticipated wind change blew the radioactive cloud toward Adelaide. The future prime minister's mother, Hazel Blair, died nineteen years later after a long battle with thyroid cancer, something the British medical researcher Dick van Steenis chalks up to exposure to radioactive fallout. "[A]s a youngster in Adelaide drinking local milk," van Steenis told *Bulletin* magazine, "Tony Blair is very likely to be at risk of bone cancer himself."

A Blair spokesperson dismissed the theory, saying: "It sounds like the silly season's been going on a little bit longer than we thought." The denial did not keep the story from showing up on the world's newswires.

Heat and the Powder Keg

"I wasn't here in 1967, but I hear so many stories . . . ," Detroit Mayor Kwame Kilpatrick told the Livingston Economic Club in 2004. "I believe at that moment, something happened, and it was a dramatic, cataclysmic shift in southeastern Michigan, and we can't recover from it."

The sweltering heat of the summer of 1967 was to cast a long shadow over the Motor City. Known in some circles as the "Summer of Love," it was also a summer of great civil unrest and social change. Detroit was then the fifth-largest city in the United States, home to the nation's automotive industry and Motown Records. In the early 1960s, Detroit attracted more federal funding than any other city except New York. In 1966, Detroit had been named an "All American City" by *Look* magazine. But racial and economic tensions were creating a powder keg beneath the surface that would need only one spark to violently explode. The temperature on Sunday, July 23, would be the match.

Numerous studies have shown that high temperatures have an effect on human personality. Dr. Lance Workman, of the

University of Glamorgan, in Wales, found that heat affects the levels of serotonin released in the brain, which can result in increased aggression. Other research suggests that when the brain is heated, the hypothalamus, a region that regulates the body's temperature, also produces extra adrenaline. Disagreements that might cause minor annoyance on a cool day have a way of escalating when the mercury rises. Most riots in the United States have occurred when the temperature was between 75° and 89°F (23° and 31°C)—warm enough to increase tensions, but not hot enough to make people too lethargic to be bothered with fighting.

The unseasonably hot summer of 1967 set off a spate of racially charged riots across the country, including 164 incidents in such cities as Cleveland and Newark; but none would be as devastating in its long-term effects as the five-day siege in Detroit that resulted in forty-three deaths, seventy-three hundred arrests, and property damage of sixty million dollars. More than four hundred buildings were destroyed citywide. The events only magnified the social problems that had fueled the racial and economic rift and caused it to widen.

Detroit, in the words of author Albert Lee, was "a town with grease under its fingernails and the distinct odor of machine oil on its breath." It is a city defined by the automobile industry, which attracted immigrants, African Americans, and other ethnic minorities in the 1940s for plentiful factory jobs.

While Detroit welcomed the factory labor, white Detroiters did not necessarily welcome blacks into their areas. Neighborhoods were still racially segregated by official ordinances until May 3, 1948, when the U.S. Supreme Court ruled that racial covenants in housing were unconstitutional. Interestingly the plaintiffs in the case were not African Americans. It was a white couple who sued their black neighbors, claiming they had no right to live next door because their deed carried a "whites only" clause.

When the Supreme Court ruled against the racial restrictions, the *Detroit Free Press* tried to quell white fears with a headline reading: MIGRATION OF MINORITIES WITHIN DETROIT DOUBTED.

Ironically, the Motor City would be hard-hit by the forces created by the automotive revolution. Freeway construction in the 1950s and 1960s (major Detroit arteries were named for Chrysler, Ford, and even the labor leader Walter Reuther) forever changed the makeup of Detroit. Effective mass transit was never fully developed in a city that wanted to encourage car ownership. Thus the roads were heavily congested, and suburban life seemed like an increasingly appealing alternative to many middle-class Detroiters.

The construction of the major I-75 highway cut through the heart of the city's African American community known as Black Bottom. (The "Black" in the name had nothing to do with African American. It had been given that name in the nineteenth century for its rich soil.) Displaced residents headed for a majority white neighborhood in the Twelfth Street section of the city. Although they could officially have moved anywhere, unofficial neighborhood policies made it clear that there were still some places where blacks were not welcome. The Twelfth Street neighborhood was one area where African Americans felt they could go. "Psychologically, black folk really said, 'Well, we can't move anywhere else,' so we continued to stack on top of each other," Detroiter Lewis Colson told the *Michigan Citizen*. As more blacks moved in, whites moved out. During the 1950s the white population of Detroit declined by 23 percent; by 1967, whites made up 70 percent of the city's population.

That said, the city was, in fact, somewhat less segregated than other major cities on average. In fact, when other cities had problems with racial violence, Detroit was seen as something of a model for racial peace. Social scientists visited Detroit to learn how urban areas could better deal with their racial problems. In March 1967 *Newsweek* magazine listed Detroit's mayor, Jerome P. Cavanaugh, at the top of a list of mayors who might have national political futures, along with New York's John Lindsay. It was said that on a clear day, Cavanaugh could see the White House. The Department of Justice's Office of Law Enforcement Assistance designated Detroit as the nation's model for police-community relations. But Detroit's African Americans were not comparing their lot to the blacks of other major cities; they were looking at their neighbors—Detroit's white residents—and many were not pleased with what they saw.

According to official reports, Sunday, July 23, 1967, was "exceedingly hot and muggy"; the daytime temperatures had been "in the nineties." That night the Detroit police raided a blind pig—an illegal after-hours bar—on Twelfth Street. Blind pigs had come into existence in the bad old days, when restaurants and bars would not serve blacks. After 1948 the law changed, but habits did not. The pigs were a source of tension between middle- and working-class African Americans. The owners of legal establishments, who had to pay for liquor licenses and the like, saw them as detrimental to their businesses and regularly complained about them. The working-class blacks who frequented the pigs tended to see raids by the predominantly white police force as an attempt to impose their social values on them.

Instead of making a few arrests while closing the place down, the police arrested everyone there—eighty-two people—and held them outside in the hot, humid air while waiting for reinforcements. Had it been a cooler evening, perhaps cooler heads would have prevailed. But this was the kind of sweltering night when it was hard to sleep in overcrowded apartments without air-conditioning. That meant that more people than usual were awake at 4:00 A.M. to hear the commotion on the street through their open windows. A crowd gathered and grew to several hundred: the mood got uglier. The violence began with shattering glass—depending on what source you believe, it was either the window of a police car or a shop. What is certain is how quickly the violence escalated. Looting and fires spread throughout the city, eventually burning through fourteen blocks. Although the Detroit police—aware of uprisings in other cities—had planned for just such an emergency, they were unable to mobilize as they would have liked, in part because on a hot summer weekend many police officers were away from the city enjoying the great outdoors.

The situation rapidly got out of control. Police were overwhelmed and federal troops were called in—the first time federal troops had been sent to contain an urban uprising. By Friday the city was patrolled by forty-four hundred Detroit police, eight thousand National Guardsmen, forty-seven hundred federal troops, and three hundred and sixty state police, assisted by Canadian firefighters from Windsor, Ontario. When the smoke cleared, forty-three people were dead—thirty of whom were killed by police or the military—seventy-three hundred were arrested and twenty-seven hundred businesses were looted. One looter who was asked by a reporter how he liked his new television said, "Not so good. The first thing I saw on it was me stealing the damn thing."

In the wake of the riots, the "white flight" from Detroit to the suburbs (in which many blacks also partcipated) accelerated at a frantic pace. Businesses and their owners fled the city. Detroit lost 36 percent of its jobs between 1970 and 1990. Almost none of the merchants in the Twelfth Street corridor returned. The Department of Housing and Urban Development repossessed tens of thousands of abandoned houses in Detroit in the 1970s—by 1976 HUD owned seventeen thousand buildings in the city, more than it did in New York, Chicago, and Los Angeles combined. The population plummeted from 1.6 million to 1 million.

Although a wave of new development in the city began in the early 1990s, Detroit has a long way to go to overcome the effects of that hot day in July. As Sidney Fine, a professor of history at the University of Michigan, told *U.S. News and World Report*, "If someone could come to Detroit today and re-create the conditions of 1967, he'd be considered a miracle worker."

Making Monsoons

"Never in the history of warfare have weather decisions played such an important role in operational planning as they did in Southeast Asia," wrote Gen. Creighton Abrams in 1968. "Khe Sanh, the A Shau Valley, and Kham Doc are only a few of the many areas where weather has been a primary consideration in operational intelligence planning."

When the United States sent troops into Vietnam, they were aware of the French experience in Indochina, but they believed that they would not run into the same problems. After all, the U.S. military was one of the strongest in the world and had superior technology. They had the best researchers and scientists in the Defense Department and a team of meteorologists on the case to make sure that everything would be timed so battles would be fought in the best weather conditions.

Not content to simply forecast the weather, strategists for the first time attempted large-scale weather control as a military tactic. Yet in the end, the United States was to discover that even with all its technological advances, nature was still in the driver's seat.

Trying to control the weather is nothing new. Some of the earliest attempts involved the use of cannons and church bells. The scientific explanation went like this: rain inevitably follows thunder; therefore loud noise must cause rain. Somehow this noise-and-rain correlation led to two contradictory beliefs: one, that warfare caused rain, and the other, that storms could be broken up by firing guns into them.

In 1871 Edward Powers, a Chicago civil engineer, published the book *War and the Weather, or, The Artificial Production of Rain*. He urged the government to fund research into "a well-defined method of causing rain to fall at will," namely, a battery of cannons set in lines facing each other to create warm and cold air currents.

In the late nineteenth century the Weather Bureau and the Department of Agriculture funded various tests in which explosives were fired into the clouds to release their rain—money that the *Chicago Times* believed would have been "less ridiculously employed if it were devoted to the attempted manufacture of whistles out of pigs' tails."

In 1916 the city of San Diego was suffering from a drought, and they turned to a famous rainmaker, who attracted clouds with a foul-smelling mix of chemicals. He had mixed up a vat of the stuff in Los Angeles and it had rained there. When he repeated his trick in San Diego, he was a bit *too* successful. Not only did it rain, it poured—the storm wiped out crops and killed people. Not only didn't Charles Mallory Hatfield get paid; he was forced to run for his life.

In 1946 a high school dropout named Vincent Schaefer, who was employed by General Electric, flew above the Schenectady, New York, clouds and dropped six pounds of dry ice into them, producing snow. A year later, Schaefer and Irving Langmuir, a

Nobel laureate in chemistry, tried to change the course of a hurricane that threatened the Florida coast. They dumped almost two hundred pounds of dry ice into its eye, and as they predicted, the storm changed direction. The experiment would have been a complete success were it not for the fact that the new path put the storm on a collision course with Savannah, Georgia, where it did about five million dollars in damage.

In any case, these attempts were much more promising than the whole cannon fiasco. By the early 1950s, cloud seeding had become commonplace—commercial seeders operated all over the drought-stricken areas of the West. The most common chemicals used for cloud seeding include silver iodide and dry ice. Private weather companies with names like Atmospherics and Better Weather Incorporated were born, and military strategists were quick to suggest adding weather modification to the military arsenal. In 1957 an advisory committee to President Eisenhower suggested that weather control could become a "more important weapon than the atom bomb."

Of course, American generals were not the only ones to think of this. In Soviet Russia scientists were looking for ways to modify weather using laser technology. Recently declassified documents in England reveal that an August 1952 storm that dropped nine inches of rain in only twenty-one hours followed a secret rain-making experiment in which the military seeded clouds with silver iodide powder. The storm created a flood that swept through the coastal town of Lynmouth, destroying bridges, buildings, and roads and killing thirty-four people.

Another recently declassified report describes the United States' future plans for weather control as envisioned in 1996. The report, "Weather as a Force Multiplier: Owning the Weather in 2025," describes cloud seeding with heat-absorbing dust, triggering lightning by using nanotechnology, and using lasers to clear fog and clouds.

That was all theory, but in 1966, during the Vietnam War, the United States had the opportunity to put some of its ideas into practice with the top-secret Project Popeye. The target of the operation was the Truong Son Strategic Supply Route, as it was called by the Vietnamese, or the Ho Chi Minh Trail, as the Americans called it. The trail, which snaked through Laos, Cambodia, and North Vietnam, was the main supply route for troops and supplies to North Vietnam. It was not what you would call a superhighway. In 1965, when Project Popeye was still in the concept stage, the trail was a narrow dirt road with a series of swaying bamboo bridges that Indiana Jones would think twice about crossing. The whole route was mostly impassable during the wet monsoon season.

The North Vietnamese recognized the importance of the trail, and as the Americans were looking for ways to destroy it—bombing was doing a decent job, but supplies were still getting through—the North Vietnamese were working to improve it. They labored day and night to cover the surface with rocks, stones, or logs, and they devised a drainage system to keep it reasonably dry in the rainy season. Each time the Americans bombed, the North Vietnamese filled the craters, and the supplies just kept on rolling.

Having no luck with the conventional means of destroying a road, the Americans became more creative, with Project Popeye, whose goal was to lengthen the monsoon season.

From 1967 to 1968, more than twelve hundred weather modification missions were flown over the Ho Chi Minh Trail. WC-130s dropped silver or lead iodide flares, which they called "olive oil," into the atmosphere to seed the clouds, and the rainfall in targeted areas increased by around 30 percent. The weather modification unit also sprayed salt from cargo planes to suppress the ground fog on U.S. runways, a practice that has become commonplace at airports.

Whether or not these attempts were successful depends on whose report you read. Successful or not, most analysts do not believe they played a significant role in the outcome of the war.

"When you're slogging through ankle-deep mud," wrote Jim Wilson in *Popular Mechanics*, "another inch of it probably doesn't make that much of a difference."

Throughout the Vietnam War soldiers battled the weather and the terrain as frequently as they battled the enemy. Mist and fog grounded air support and supplies. During the rainy season, soaked soldiers slogged through mud, but the "dry" season was "dry" only by comparison. The rain did not fall from the sky, but humidity hung in the air, as in a steam room. The soldiers took a daily regimen of pills but still suffered from diarrhea and malaria as well as heat stroke, dehydration, and exhaustion.

One of the most curious "weather" phenomena encountered by U.S. soldiers stationed in Vietnam was a sticky, yellow rain. The soldiers ran for their gas masks, assuming it was a chemical attack. It turns out it was a bombardment of giant honey bee droppings. A tropical bee species, *Apis dorsata,* deals with extreme heat by making flights over the jungle and defecating en masse, then flying home. For some reason, this keeps them from overheating the colony.

Weather played a significant role in the battle of Khe Sanh. Like Dien Bien Phu, the base at Khe Sanh was located in a remote, isolated area in the mountainous highlands. It was defended by 6,680 marines who needed about 235 tons of food and supplies per day, so air supply was absolutely critical. U.S. Command was quite familiar with the poor weather conditions to be expected during the northeast monsoon. Low ceilings and poor visibility were to be expected, but the commanders still felt that given the average conditions everything would work out fine.

In the winter of 1968, however, conditions at Khe Sanh were not average. The cloud cover was much thicker than normal, with fog reducing visibility to less than five miles most days. The poor flying conditions threatened the marines' lifeline. "On many a morning when visibility was excellent, the runway remained shrouded in mist," noted one official report. "A deep ravine at the east end of the runway seemed responsible, channeling warm, moist air from the lowlands onto the plateau where it encountered the cool air, became chilled, and created fog." The marines dubbed this ravine "the fog factory." Combined with North Vietnamese antiaircraft fire, the conditions paralyzed most supply traffic.

The main battle for Khe Sanh began on January 21, 1968. North Vietnamese artillery pounded the camp, but the marines were unable to locate the source because of the morning fog. The runway was damaged, several helicopters were destroyed, fuel-storage areas went up in flames, and the main ammunition dump of fifteen hundred tons of munitions exploded.

The siege was appearing all too familiar. "I don't want any damn Dien Bien Phu," President Lyndon Johnson said. "A newspaper article went on to say that the President had received assurances from the Joint Chiefs of Staff that the base would not fall," wrote R. D. Camp Jr, a veteran of Khe Sanh, who added, "No one had asked our opinion."

After the initial barrage the NVA and Viet Cong forces continued to bombard Khe Sanh daily. Landing supply planes became too hazardous, and the air force changed its strategy—dropping cargo pallets out of an aircraft using parachutes. Overcast skies precluded the use of even this system most of the time. Delicate cargo, like medical supplies, could not be transported this way.

"With Khe Sanh surrounded by an enemy estimated at between twenty thousand to forty thousand all eyes turned toward the sky: the Americans for resupply, the North Vietnamese for the B-52 gunships and tactical air bombardments that rained death from above," wrote military historian Harold A. Winters in *Battling the Elements*. "For both, the weather held the key."

Throughout February and March the battle of the clouds continued with air support at times grounded by fog, and at times permitting resupply. As the skies cleared toward the end of March, Operation Pegasus was planned. The mission was timed to coincide with the breakdown of the northeast monsoon season and designed to relieve the siege. On April 1, combined army, marine, and Vietnamese forces marched toward Khe Sanh through moderate resistance. They arrived on April 2 and battled until April 14. The next day, the operation officially ended. Khe Sanh was a tactical failure but a psychological victory in that most of the American soldiers escaped the fate of the French at Dien Bien Phu.

"In hindthought it can be truly stated that the tropical jungle presented one of the most difficult environments for combat," wrote Erik Durschmied, who covered the war as a television reporter. "Intense heat and continuous humidity broke down the built-in defenses of the body. The rain bred clouds of deadly insects. . . . [T]he elements, more than the black-pyjamaed battalions, were the enemy. In the jungles and in the paddies of Southeast Asia, nature was in control."

Lucy and Her Friends

The entire field of archaeology owes a debt to Mother Nature. The same flash floods and storms that bury and preserve bones and fossils uncover them centuries later, giving scientists and historians a new window into the past. While many archaeological discoveries are the result of painstaking research, some involve a great deal of luck—and fortunate weather.

Take for example Lucy. In November 1974 anthropologist Donald Johanson and his graduate student, Tom Gray, were fossil hunting in Hadar, Ethiopia. As they searched along a gully, they spotted a bone sticking out of some soil that had been eroded by a recent flash flood. The bone, hidden for millions of years in sediment and volcanic ash, was just the beginning. After three weeks of excavation, Johanson and his team found several hundred pieces of bone, all belonging to one hominid female. The scientists gave her the nickname Lucy after the Beatles song "Lucy in the Sky with Diamonds," but her official name is *Australopithecus afarenis.*

The Lucy skeleton was dated back about three million years,

the most complete and oldest hominid skeleton to have been found at that time. Lucy was about a meter tall and walked erect, which opened new debate in the scientific community as to when and why humans started standing upright.

"If I had waited another few years," Johanson wrote in his book *Lucy: The Beginnings of Humankind*, "the next rains might have washed many of her bones down the gully. . . . What was utterly fantastic was that she had come to the surface so recently, probably in the last year or two. Five years earlier, she would still have been buried. Five years later, she would have been gone."

Violent winter storms also increased our knowledge of Native American history and culture. For years, Richard Daugherty of Washington State University had been excavating buried remains at the abandoned coastal village of Ozette. Little by little he tried to piece together the history of the Makah Indians who had once lived on the land. Their descendants told him a story of a huge mudslide that had buried the entire village. Daugherty was unable to confirm the story until 1970, when nature lent a hand. A storm sent tides raging up the beach at Ozette and washed away a bank.

Under the soil was a vast deposit of artifacts dating from around the time of Columbus's arrival in the New World. There were fishhooks of wood and bone, a harpoon shaft, a canoe paddle, a woven hat, and parts of inlaid boxes. The objects are now housed in a museum created by the Makah Tribal Council.

Most fortuitous of all was the accidental discovery of a Stone Age body by two nonscientists. German hikers Helmut and Erika Simon were walking in the mountains of the Austrian-Italian border on a particularly sunny morning in September 1991. They wandered off the trail and were startled to see a dead body in the melting ice. They assumed it was the corpse of a modern climber, and they called a rescue team. As the medics chipped him from the ice and unearthed many of his belongings, it became clear that this climber was quite a bit older than they had originally suspected.

In fact, the man's body dated from 3300 BC. It had been preserved in ice until a fall of dust from the Sahara and an unusually warm spell melted the ice in 1991, bringing the "Ice Man" back to the surface. Since he was found in the Ötztal Alps, he was given the nickname Ötzi. What was unique about Ötzi was that he was incredibly well preserved, with his clothing and tools intact. This gave scholars a greater understanding of Ötzi's life, culture, and community.

A unique set of circumstances gave us this window into the past. Soon after Ötzi died, his body had been covered with snow, which kept the predators away. Then a glacier covered him and entombed him in ice. Normally, a glacier would destroy everything in its path, but Ötzi's body was sheltered in a rock hollow.

Even more amazing was that a freak thaw unearthed him just as a pair of hikers was passing. As one commentator wrote, "Over the past five thousand years the chance of finding the Ice Man existed for only six days."

Operation Thwarted by Desert Storm

It is called *haboob,* a name derived from the Arabic word for "phenomenon." It is a dust storm that travels as fast as 80 kph (50 mph). The downdrafts from nearby thunderstorms pick up sand and create a wall of grit as high as 914 m (3,000 ft). Near the source of the storm, visibility is close to zero. The sand-haze remains in the air for days, and when it finally settles, it coats everything in its path.

Haboobs occur in desert regions in many parts of the world, especially in Saharan Africa and the Middle East. The Zagros Mountains that separate Iran and Iraq produce enough lift to create convective storms that spread over the Persian Gulf, especially along the Iranian coast. They are more difficult to forecast than many other types of storms—a fact that was to prove disastrous to a secret mission by the United States to rescue American citizens held hostage in Iran.

At 10:30 A.M. on November 4, 1979, a mob of three thousand armed students stormed the U.S. Embassy in Tehran and took the sixty-six occupants hostage. They would eventually release a few women and African Americans, leaving fifty-two Americans in captivity. The students demanded that the United States return Reza Shah Pahlavi, who had left Iran in poor health in January, to stand trial. President Jimmy Carter had allowed him to enter the United States for medical treatment.

The shah had been the autocratic ruler of Iran from 1941 to 1979, with one brief exception in 1953 when he was briefly overthrown by Mohammad Mossadeq, the prime minister. His return to power was accomplished with the assistance of Britain's MI6 and the CIA. Why did the Western powers want an autocrat to remain at the helm? Because the shah was pro-Western, and anti-Communist—and on top of that, Iran had oil.

The shah was credited with greatly modernizing Iran and introducing many reforms that pleased his Western allies but angered fundamentalist religious leaders. He imposed his reforms on the populace by arresting anyone who spoke out against them. He spent large sums of money on military hardware—by 1979 Iran was spending four billion dollars a year on U.S. weapons. When a spike in oil prices in the mid-1970s caused rapid inflation in Iran, the shah's lavish spending became ever more apparent in contrast. He had several royal palaces and numerous overseas properties; he lined officials' pockets by awarding them lucrative defense contracts.

He was increasingly unpopular, not only among fundamentalists, but also with leftists and secular nationalists. They all agreed that they wanted the shah—and the Western powers that influenced him—out. The opposition, led by Ayatollah Khomeini, drove the shah out in 1979. Now in power, the ayatollah made it his goal to expel Western influence and create an Islamic state in Iran. He was able to galvanize the various factions in the country by uniting them against a common enemy, the United States.

This led directly to the capture of the American hostages on November 4. Initially President Carter tried to get the hostages back through negotiations, but the talks broke down in April 1980. The only option, it appeared, was military intervention. An ambitious rescue, code-named Operation Eagle Claw, was set in motion.

Tehran would not be an easy site for a rescue. It was surrounded by seven hundred miles of desert and mountains in every direction; thus, a multistage rescue was needed. The mission was to take place over two nights. A rescue team would pilot eight helicopters from a staging area in Oman to a region about fifty miles outside Tehran, called Desert One. Along with the helicopters, there would be a team of C-130 fixed-wing airplanes. These would fly the troopers and the reserve fuel to Desert One.

The next day, American intelligence agents would rendezvous with the troopers and take them to the embassy in a truck. Once they had the hostages, they would move them to a nearby soccer stadium. The helicopters would pick up the Americans and evacuate them to Manzariyeh Air Base. There was no room for error. Fortunately, the forecasters said the weather on the mission date, April 24, would be clear, with a full moon providing ample visibility. Unfortunately, they were wrong.

The mission got off to a bad start when, two hours into the flight, one of the eight helicopters was grounded by an equipment problem. The pilots were picked up by one of the other helicopters, which was now running behind the group by about fifteen minutes. The remaining six, flying five hundred feet above Iran's Great Salt Desert, came face-to-face with a haboob, which decreased visibility to one mile. It was, as one pilot described it, "like flying in a bowl of milk."

The pilots, under strict radio silence, had planned to maintain visual contact with one another, but it was utterly impossible. The helicopters, which had been moving as a team, were now operating independently.

Just when they thought they were past the worst of it, they were socked with a second dust storm. At this point, the attitude indicator on one of the helicopters failed. The pilot, flying blind without the equipment that would tell him if he was right side up or upside down, became disoriented and dizzy. He decided to head back. Ironically, had he continued he would have cleared the dust storm in about twenty minutes. Instead, he had to go the entire distance back through the two clouds of dust. There were only six rescue helicopters left.

The remaining choppers started arriving at the site, one by one, each between fifty and ninety minutes late. At Desert One, yet another helicopter experienced equipment failure. With only five working helicopters, it was decided that the mission could not safely continue. After refueling, the helicopters started back; but one kicked up a blinding dust cloud, then collided with a C-130—both were destroyed, and eight Americans were killed.

The disastrous mission had far-reaching consequences. It reinforced Ronald Reagan's charge that the Democrats had allowed the country's military to deteriorate. By the time of the election of 1980, the American hostages had been held in Iran for a year. Frustration over the hostage crisis influenced voting. On election day, Reagan ousted Carter with 51 percent of the popular vote. The day after his inauguration, the new president announced that Iran had agreed to release the remaining American hostages. The timing led to suspicions, which were never substantiated, that the Reagan campaign had made some sort of secret deal with the Iranians to prevent Carter from winning the release of the hostages before the election. A more likely scenario is that the invasion of Iran by Iraq (in September 1980) lessened Iranian resolve to hold the American hostages—they had bigger problems to deal with.

Internationally, many political strategists believe that the lack of decisive and effective action by the United States in the hostage crisis offered the Soviet Union the opportunity to expand its influence and that this was a factor in the Soviet invasion of Afghanistan. The Soviets entered Afghanistan to support the People's Democratic Party of Afghanistan against the fundamentalist Muslim, anti-Communist mujahideen. The mujahideen were loosely organized, untrained soldiers, but their fighting ability improved dramatically with instruction and arms supplied by the United States. An invasion that began with thirty thousand Soviet troops ultimately grew to one hundred thousand, yet there were no signs of success as the casualties mounted. By the time the war ended, fifteen thousand Soviets and one million Afghans had died. The Afghan War is widely considered to be a major factor in the downfall of the Soviet Union. It would also cast a long shadow over Afghanistan. The various guerrilla forces that had been united against an outside enemy were not easily united with one another after the war. The political divisions set the stage for the rise of the Taliban.

Meanwhile, in the United States, still smarting from the Iranian crisis, President Reagan opened a back door to Iraq, while remaining officially neutral in the Iran-Iraq War. Under the theory that "the enemy of my enemy is my friend," the United States channeled intelligence and hundreds of millions of dollars in loan guarantees to Iraq, indirectly helping the war effort. Saddam Hussein, militarism aside, was taking his country in a secular, not fundamentalist, direction. He offered free education, encouraged women to become educated, and provided new housing and health care, which appealed to Western sensibilities.

The Iran-Iraq War dragged on for eight years, making it the longest conventional war of the twentieth century. When it finally ended in July 1988, after a United Nations–mandated cease-fire, the final death toll was estimated to be around 1.5 million, yet Iraq had not gained an inch of ground. The nation was left with huge debts. One country that had lent Iraq money was the wealthy but tiny Kuwait. By 1990 the Kuwaiti emir was getting more insistent that he be paid back. At the same time, Kuwait was flooding the oil market with surplus oil and lowering prices. Saddam Hussien decided he could end both these problems by invading Kuwait.

The invasion of Kuwait made Iraq's other neighbor, Saudi Arabia, very nervous. The Saudis accepted an American offer to protect their oil fields from Hussein's covetous glance. This stepped on the toes of a former Afghan warrior, the Saudi-born Osama bin Laden, who had wanted to provide Saudi Arabia with a force of mujahideen to defend the nation. That King Fahd would choose Western infidels over his army infuriated bin Laden. It was the main justification he gave for the attacks on the United States on September 11, 2001. Had it not been for a haboob over the Great Salt Desert one night in 1980, who knows how different the world might have been.

Meteorology and Rockerty

January 28, 1986, was especially chilly for Cape Canaveral, Florida, where the average low in January is 47°F (8°C). That morning the mercury had dipped to 28°F (–2°C), but the morning chill could do little to dampen the spirits of the teachers, students, and family members who filed onto bleachers to witness the launch of NASA's twenty-fifth shuttle mission. Their jackets were covered in buttons celebrating Christa McAuliffe, the first "teacher in space." The thirty-six-year-old had been chosen in a highly publicized search lasting ten months. More than eleven thousand applicants had been screened out in favor of the pretty, warm everywoman. McAuliffe was going to teach two classes from space and, it was hoped, by telling us what it was like up there, bring back some of the magic of the early space flights. She was to represent all of us. The unassuming educator had become an instant celebrity, and she appeared humble and excited when she was featured in *People* magazine and on CNN.

"I want to give an ordinary person's view of space, the idea that there's a new way of living out there," she said on the

MacNeil/Lehrer NewsHour. "[T]here's going to be space law, there's going to be business in space, and students have to prepare for that future."

President Ronald Reagan would even be mentioning the teacher in his State of the Union address. In a memorandum dated January 8, NASA had proposed the following language:

"Tonight while I am speaking to you, a young elementary school teacher [she was actually a high school teacher] from Concord, New Hampshire, is taking us all on the ultimate field trip as she orbits the earth as the first citizen passenger on the space shuttle. Christa McAuliffe's journey is a prelude to the journeys of other Americans living and working together in a permanently manned space station in the mid-1990s. Mrs. McAuliffe's week in space is just one of the achievements in space we have planned for the coming year."

Indeed, another thirteen shuttle missions were planned for 1986, and NASA was holding a new contest to select the first journalist in space. It was enough to make you forget that the Soviets were launching far more rockets each year than the Americans.

Until early Sunday morning, NASA officials were not sure if the *Challenger* launch would have to be delayed due to weather conditions. A cold front from Texas and even the weather in Africa threatened to hold things up. The forecast for Saturday called for a chance of rain and thunderstorms in the area at launching time. Cloudy skies forced NASA to cancel some crew flights aboard shuttle training jets that day. Although the cold front was not expected to pass through entirely until Monday, the storm clouds had moved along, and blastoff remained set for Sunday.

Thunderstorms are a serious problem in rocketry. Not only do rockets conduct electricity, they actually trigger lightning strikes. Unlike aircraft, which fly horizontally, rockets shoot straight up. They go through very rapid changes in the ambient atmospheric electrical field. The highly conductive metal skin doesn't have time to adjust to the conditions. And as they rise, rockets trail long plumes of conductive ionized gases—perfect conditions for a lightning strike.

This phenomenon was first observed in 1969 during the launch of *Apollo 12*. Although no lightning had been forecast by the Cape Kennedy Air Force station on November 14, shortly after takeoff observers saw two lightning streaks flash around the vehicle. After thirty-six seconds a brief power shutdown occurred, but the mission went off without any further hitches. After an investigation, NASA determined that the rocket had triggered the bolts. The Saturn 5 rocket was, in essence, a giant lightning rod.

"A lightning bolt occurs when the electrical charges on the earth and in a cloud differ enough to bridge the essentially non-conducting gap of air between them," a NASA press release explained. "The ascending rocket carried on its surface the same charge as the ground, and its hot exhaust plume acted as a conducting wire, down which the lightning traveled to the ground."

After this lesson, NASA safety regulations dictated that launches be scrubbed if lightning was detected within five miles of the launch pad. Yet this precaution would not be enough to prevent the loss of a seventy-eight-million-dollar Atlas-Centaur rocket in 1987. The rocket, which was carrying an eighty-three-million-dollar military communications satellite, had to be destroyed after it careened out of control. Four lightning bolts were detected about forty-eight seconds after liftoff—the electrical fields created by the rocket-triggered lightning strikes interfered with the electrical systems on the craft.

In June 1987 three rockets at NASA's Wallops Island, Virginia facility were due to be launched to study the effects of lightning on the ionosphere. The lightning, apparently, grew impatient. As the NASA team waited for a thunderstorm to pass, lightning ignited an Orion rocket and two smaller test rockets. The test rockets, intended to help the scientists check out its tracking radar, lifted off and followed their planned routes. The Orion, on the other hand, had not yet been elevated to its liftoff angle, so it shot off horizontally. Fortunately no one was hurt—the rocket flew about three hundred feet before crashing into the ocean.

Of course, getting into space is only half the journey. A manned rocket must also bring its passengers home. This is why the weather in Africa was taken into consideration when launching *Challenger.* Two emergency landing strips in Africa, including the primary overseas abort site at Dakar, Senegal, would probably be unusable at the time of the launch because of a reduction in visibility from dust blowing in from the African desert. A backup emergency field was arranged at Casablanca, Morocco, but forecasters predicted there would be thick cloud cover over the site on launch day. If both emergency landing sites were unusable, the voyage would have to be delayed. In the end, however, it was decided that conditions were satisfactory.

Some engineers from Morton Thiokol, manufacturers of *Challenger*'s solid rocket booster, were still concerned about the weather. In a teleconference the afternoon before the launch, the engineers argued that the launch should be delayed. They did not have enough data to predict how the rocket-motor seal would work in low-temperature conditions. Each of the shuttle's two solid rocket boosters was made up of four sections, each of which was held together by a tang on one section that fit into a coupler on the other. A rubber O-ring and putty made of zinc chromate filled with asbestos formed a seal at the join. A second ring acted as a backup. When the solid rocket boosters fired, the combustion of the propellant would produce tremendous heat and pressure. At the same time, the million pounds of thrust from the liquid-fuel rockets would create enormous forces and bend the solid rocket boosters slightly. Only the O-ring would keep hot gases from blowing out of the seal. *Challenger* had been sitting on its launching pad for thirty-eight days prior to launch. During that time, 17.8 cm (7 in) of rain had come down. No one was sure how these conditions would affect the rings. There was already evidence that low temperatures could have unintended and deadly consequences. During a shuttle launch the previous year, hot gases perforated one of the booster's primary O-rings. In this instance, the secondary ring kept the gas from escaping.

After much debate, managers at Morton Thiokol overrode the engineers' assessment and recommended that the liftoff take place on January 28, only slightly delayed from 9:36 A.M. to 11:38 A.M. Christa McAuliffe carried her son's toy frog as she boarded the craft along with her crewmates Dick Scobee, Michael Smith, Ronald McNair, Judy Resnik, Ellison Onizuka, and Gregory Jarvis. The temperature at launch time was still below freezing.

As Christa McAuliffe's parents gazed skyward, proudly sporting their "teacher in space" badges, the shuttle blasted off. A little more than a minute later, *Challenger* exploded. The spectators gazed into the sky, unsure whether the blast was expected or whether something had gone horribly wrong. The NASA technician reading electronic data over the public address system had not been prepared for these circumstances. "Flight controllers here looking very carefully at the situation. Obviously a major malfunction."

"The major malfunction, of course, was that the entire crew was dead," wrote *Time* magazine reporter Alan Richman.

After the disaster a commission was assembled including Neil Armstrong, the first man to walk on the moon; Sally Ride, the first American woman in space; and Richard Feynman, a Nobel Prize–winning physicist who had worked on the Manhattan

Project. They interviewed one hundred and sixty people and examined one hundred thousand pages of documents—the transcripts of their hearings filled twelve thousand pages. They determined that cold weather and a defect in the solid rocket booster were to blame for the astronauts' deaths.

Some moisture from the rain had probably gotten into the join area and frozen, reducing the resiliency of the O-rings, which could not contain the hot fumes of the propellant. A second after the solid rocket booster was ignited, camera evidence showed puffs of dark smoke coming from the joints in the right motor where the O-rings were being burned. Less than a minute later, a flame was visible. The flame was directed toward the external tank by aerodynamic forces: it penetrated the external tank and severed it. Hydrogen escaped and mixed with gases escaping from the oxygen tank, resulting in a huge explosion. The solid rocket boosters kept flying in an unguided trajectory until the U.S. Air Force safety officer destroyed them. The *Challenger* cabin was torn from the shuttle and plunged into the Atlantic Ocean. The commission determined that the crew members survived the explosion, and some may have been conscious during the accident, but none could survive when the cabin hit the ocean at 207 mph and disintegrated. Their remains would not be found until March. Christa McAuliffe was buried on a hilltop two miles from the high school where she taught.

The *Challenger* explosion could have been a much greater disaster. The next scheduled flight of the shuttle was to carry forty-seven pounds of plutonium to power a long-distance space probe. A hydrogen burn ten miles above Florida could have ruptured the lead casket that contained the plutonium and spread it up and down the coast.

Blood Rain and World War III

The Romans called it *blood rain*, the Chinese *candle dragon*, Eurasians *wind light*. The Inuit believed it was the highest level of heaven, where the dead danced. The aurora borealis, or northern lights, those waves and flashes of color in the northern sky, have fascinated observers for years. Modern meteorologists and geophysicists know it as an electromagnetic storm.

When scientists Jan Holtet of the University of Oslo and Craig Pollack of NASA came up with the idea of putting an automated laboratory in the ionosphere above the North Pole to study the aurora, little did they know they might put the world on the brink of global thermonuclear war.

With funding and technical expertise from NASA, Norway planned to launch its largest rocket ever. The Black Brant XII was a four-stage rocket, twice as large as anything fired by Norway in the past. It was designed to travel along an arcing ballistic trajectory for fifteen hundred kilometers.

"Weather decides when a launch can be made," Ingvard Havnen of Norway's Foreign Ministry would later say as he explained why Norway could not give international airmen and missile watchers an exact time for the planned launch of the scientific missile. They did, however, have it narrowed down to a particular window—it would be fired sometime between January 15 and February 5, between 5:00 A.M. and 12:00 A.M. This is what the Norwegian rocketry director told Russian authorities. But you know how sometimes someone gives you a message and you forget to pass it along? That is what happened in this case. Word of the Norwegian launch failed to reach the Russian military's general staff.

When the rocket blasted off at 6:24 A.M. Zulu (GMT) time on January 25, 1995, Russia's Missile Attack Warning System detected it. The unexpected launch appeared to have its origin in the Norwegian Sea, where U.S. submarines were known to patrol. Was it possible that this was a surprise attack of some kind? With no more information to go on, the commander of the warning system did what he was trained to do: he treated it as a real threat. Within a matter of minutes the military officer in charge of Boris Yeltsin's nuclear briefcase saw a warning light flashing. The screen inside the case showed a possible nuclear missile with an unknown target. At 6:28 a teletype sprang to life at missile command of the Strategic Rocket Forces. "Nuclear alert. Not a drill." President Boris Yeltsin had the nuclear briefcase open and the ability to launch forty-seven hundred strategic warheads at the press of a button, something that had not happened in the course of the entire cold war. It was the first time in history that the "nuclear briefcase" was switched into alert mode, giving one man this awesome ability.

If this was a U.S. attack, it would surely be followed by a massive launch of submarine-based missiles. If Russia was going to strike back, it would have to be now, before the electromagnetic pulse of dozens of nuclear strikes rendered all their electronic equipment inoperable. But why would the United States attack Russia now, four years after the fall of the Soviet Union? If Yeltsin made the wrong decision, the consequences would be astronomical.

Yeltsin contemplated aiming an antimissile at the incoming rocket, but if it truly was a nuclear missile, shooting it down could just spread the payload farther than if it was allowed to land. What is more, an electromagnetic pulse near the magnetic North Pole would actually have a greater effect. After a few moments, the Black Brant's trajectory shifted away from Moscow, and it became clear that it would not land on Russian soil. Boris Yeltsin and Russia's top military brass continued to watch the missile for the remainder of its twenty-four-minute journey until it landed in the Spitsbergen archipelago in the Arctic Ocean.

The next day Yeltsin, still shaken from this close call, told an audience, "I have indeed used yesterday for the first time my 'little black case' with a button that is always carried with me."

Because of cold war secrecy, this is the best-known nuclear close call, but it was not the only one. According to the report "Narrative Summaries of Accidents Involving U.S. Nuclear Weapons, 1950–1980," published by the U.S. Defense Department, there were thirty-two accidents classified as "Broken Arrows"—nuclear devices lost, burned, dropped, or accidentally detonated. There are also lesser categories of nuclear accidents including "Bent Spears," "Dull Swords," and "Empty Quivers."

On at least four occasions in 1961 and 1962, U.S. Jupiter missiles with 1.4-megaton nuclear warheads were struck by lightning at their bases in Italy. The bolts activated the thermal batteries, and in two cases, the weapons were partially armed. After the fourth such incident the air force added lightning strike diversion tower arrays to its Italian and Turkish missile launch sites.

In 1979, in a moment that seemed to come straight from the

movie *War Games*, a training program was fed into the NORAD early warning system and was mistaken for an actual large-scale nuclear attack. In 1983 a solar storm tricked the Soviet early warning satellites into believing the United States was launching a massive attack.

According to the Brookings Institution, despite the ease in cold war tensions between the United States and Russia, each side is still prepared to launch its missiles at the other in minutes. The full time needed for preparation and launch of U.S. warheads is twenty-two minutes; for the Russians, thirteen minutes. (The reason for the discrepancy is that the Russian timetable assumes a missile attack from U.S. submarines, while the American timetable is based on a long-range attack from Russian-based missiles.)

Fortunately, experts believe that a true accidental launch would be unlikely. No matter what the equipment says, the people at the helm are not likely to believe that the other side is launching a massive unprovoked first strike. That is, as long as a glitch doesn't happen to occur coincidentally during a time of heightened political tensions.

Sleep well.

Nature Does Not Carry a Passport

History has demonstrated time and time again one simple truth: nature does not carry a passport. It rains down on everyone and everything with equal unconcern—the rich as well as the poor, the mighty as well as the small. In the immortal words of the poet E. E. Cummings, "the snow doesn't give a soft white damn whom it touches."

In a world of nations and states, weather is the great equalizer, reminding us, from time to time, that we share one world and the air knows no boundaries. We inhabit a small planet where soil erosion in Africa affects the rainfall in Australia, where the sun reflecting off the New York pavement causes rain in Uzbekistan. Storm systems have never recognized superpowers, and they show no sign of starting now.

With our radar and air-conditioning, our satellites and superconductors, we are still, and probably always will be, at the mercy of the elements. Despite all our noble (and ignoble) attempts at weather modification, the interconnected system is still too complex for us to control—at least not intentionally; at least not well.

As the United States learned in its experiments with redirecting hurricanes and seeding monsoons, what works on a small scale often creates unexpected havoc. Try to control a storm system in one spot, and it pops up somewhere else, like a mechanical creature in a Whac-a-Mole arcade game.

Yet nature does respond to us, whether we intend it or not. The conveniences of modern life—many of which were designed to shield us from the elements—influence the environment and the weather. In urban areas, where greenery is scarce and pavement is not, tall buildings block the path of winds and expand the surface areas that absorb solar heat. The result is an effect known as the "urban heat island." It is particularly pronounced in Japanese cities like Tokyo, where high humidity multiplies the effect of rising heat. Tokyo today is 3°F hotter than it was a century ago. Palm trees native to subtropical southern China are springing up in the city as flocks of parakeets native to southern India and Sri Lanka fly overhead. NASA scientists observing satellite images of Atlanta, Georgia (nicknamed "Hotlanta" for its nightlife), found that the hottest parts of downtown were as much as 10° F hotter than the surrounding area and that this difference caused air to rise, creating thunderstorms. If you live east of the city and a tornado comes your way, you may have Hotlanta to thank.

Try to solve the heat problem by switching on the air-conditioning and you only make things worse. Air conditioners produce hot air as a byproduct of the cooling process, and the waste heat is released outdoors. New research suggests that the waste heat can add as much as two degrees (C) to outdoor urban temperatures. Who said nature didn't have a sense of humor?

Here's another one: all those snowbirds who move to Florida to escape northern cold snaps may actually be causing freezing. In the early 1900s, as citrus growers moved to areas less prone to freezes, they drained wetlands, diverted rivers, and unexpectedly changed the climate. When wetlands disappear, temperatures drop by a few degrees Celsius. The wetlands provided a buffer that absorbed heat during the day and released it at night.

If all of this history and science has taught us anything it is this: We are neither the masters of the weather nor the servants of it—we are in a marriage with it. It is a union that encompasses all the people and all the creatures of the earth. We can try to ignore this vital fact, but we do so at our own peril, for the weather will continue to be the weather, whether we like it or not.

Bibliography

Aarkrog, Asker. "Global Radioecological Impact of Nuclear Activities in the Former Soviet Union," Proc. International Symposium Impact of Radioactive Releases, IAEA-SM-339/128, Vienna (1995).

Adams, Douglas. *The Salmon of Doubt*. New York: Harmony Books, 2002.

Adams, Jad. "Who Killed Kitchener?" *The Sunday Telegraph*, June 7, 1998.

Adkins, Lesley, and Roy A. Adkins. *Handbook to Life in Ancient Greece*. New York: Facts on File, 1997.

Agence France Press. "Disaster, Screams and Sunsets, Krakatoa's Legacy Recalled by Asian Tsunami." January 4, 2005.

Agence France Press. "50ème anniversaire de la bataille de Dien Bien Phu, le 7 mai 1954." May 6, 2004.

Agence France Press. "Reagan Played Decisive Role in Saddam Hussein's Surival in Iran-Iraq War." June 9, 2004.

Ahearn, Lorraine. "Cold War: Marines Get History Lesson." *News and Record* (Piedmont Triad, NC), January 27, 1996.

Alperovitz, Gar. "Hiroshima: Historians Reassess." *Foreign Policy*, June 22, 1995.

American Journalism Review. "Newseum Nugget." (June 1999).

Anderson, Scott. "Bloody Battle Costs Burnside His Command." *Washington Times*, January 10, 1998.

AP Worldstream. "Scientist Says British Prime Minister's Health Possibly Jeopardized by Atomic Bomb Test." August 31, 2004.

AP Online. "Richmond, VA Honors Slave." October 29, 2002.

Aptheker, Herbert. *American Negro Slave Revolts*. New York: International Publishers, 1963.

Associated Press. "Bell Buys Stradivarius for $4M." October 29, 2001.

Atwood, Margaret. "Napoleon's Blunders Show Risk of Pre-Emptive Strikes." *Cincinnati Post*, March 20, 2003.

Authentic History Center. http://www.authentichistory.com.

Bain, R. Nisbet. *Charles XII and the Collapse of the Swedish Empire, 1682–1719*. New York: G. P. Putnam's Sons, 1895.

Ballard, Robert D. "Deep Black Sea." *National Geographic* (May 2001).

Barnard, Bryn. *Dangerous Planet: Natural Disasters That Changed History*. New York: Crown, 2003.

Barrett, Michael. "South: The Race to the Pole." *History Today* (October 2000).

Barry, William. *The Papacy and Modern Times: A Political Sketch, 1303–1870*. New York: Henry Holt and Company, 1911.

BBC News. "Science Battles for Scott's Reputation." September 10, 2001.

BBC News. "Scott Caught Out by Cold Snap." November 9, 1999.

BBC Online. "Historic Figures: Lord Kitchener of Khartoum." http://www.bbc.co.uk/history/histocir_figures_lord.shtml.

Beauford, Fred. "American Slavery: The Ties that Bind." *Black Issues Book Review* (January 2005).

Beevor, Antony. *Stalingrad: The Fateful Siege: 1942–1943*. New York: Viking, 1998.

Behringer, Wolfgang. "Climatic Change and Witch-Hunting: The Impact of the Little Ice Age on Mentalities." *Climate Change* 43, no. 1 (1999).

Bennett, Will. "'My God, what have we done?' Hiroshima Logbook for Sale." *The Daily Telegraph*, March 26, 2002.

Best, Gary Dean. *The Nickel and Dime Decade: American Popular Culture During the 1930s*. Westport, CT: Praeger Publishers, 1993.

Biel, Steven, ed. *American Disasters*. New York: New York University Press, 2001.

Boller, Paul F. *Presidential Campaigns*. New York: Oxford University Press, 1984.

Boot, Max. *The Savage Wars of Peace: Small Wars and the Rise of American Power*. New York: Basic Books, 2002.

Borenstein, Seth. "Air Conditioners Heating Up U.S. Cities." Knight Ridder, July 17, 2002.

Bothwell, John. "'All Is at Stake' at Salamis." *Naval History* (February 2005).

Boyd, Andrew. "Wolfe Tone: Republican Hero or Whig Opportunist?" *History Today*, June 1, 1998.

Bozich, Stanley J., and Jon R. Bozich. *"Detroit's Own" Polar Bears: The American North Russian Expeditionary Forces, 1918–1919*. Frankenmuth, MI: Polar Bear Publishing Co., 1985.

Bridges, Peter. "Gripping Account of Fredericksburg's 'Winter War.'" *Washington Times*, January 11, 2003.

Brinton, Crane, John B. Christopher, Robert Lee Wolfe. *A History of Civilization*, 2nd ed. Englewood Cliffs, NJ: Prentice Hall, 1960.

Brown, Dale Mackenzie. "The Fate of Greenland's Vikings." *Archaeology*, February 28, 2000.

Bryant, Michelle. "Witch Trials." http://www.utexas.edu/features/archive/2004/witches.html.

Bryson, Bill. *The Mother Tongue: English and How It Got That Way*. New York: William Marrow, 1990.

———. *In a Sunburned Country*. New York: Broadway, 2000.

Camp, R. D., Jr. "Memories of Khe Sanh." *Naval History* 18, no.1 (2004).

Carey, John, ed. *Eyewitness to History*. New York: Avon Books, 1990.

Carnes, Mark, ed. *Past Imperfect: History According to the Movies*. New York: Henry Holt and Company, 1996.

Cavendish, Richard. "The Birth of Richard III." *History Today* (October 2002).

———. "Boniface VIII's Bull Unam Sanctam." *History Today* (November, 2002).

———. "The Crimean War Begins." *History Today* (March 2004).

———. "The Fall of Dien Bien Phu." *History Today* (May, 2004).

Cerveny, Randy. "The Oddities, Lightning." *Weatherwise* (March 1, 2004).

———. "Power of the Gods: Ancient Cultures Were Grounded on Fear of Lightning." *Weatherwise* (April 1994).

Chadwick, Alex. "Analysis: Iranian Expatriates Discuss the Iranian Revolution of 25 Years Ago." National Public Radio Special, January 16, 2004.

Chapman, James. "The Rainmakers: Large Buildings Alter the Climate in Our Big Cities." *The Daily Mail*, September 13, 2002.

Chivers, C. J. "Hunting Nuclear Waste Dumped in Moscow: Fallout of Arms Race Hits Close to Home." *International Herald Tribune*, August 11, 2004.

Clark, Stuart. *Thinking with Demons: The Idea of Witchcraft in Early Modern Europe*. Oxford: Oxford University Press, 1999.

Clarke, Philip. *Where the Ancestors Walked: Australia as an Aboriginal Landscape*. Crows Nest, NSW (Australia): Allen & Unwin, 2003.

CNN. "Cool Weather May Be Stradivarius' Secret." December 8, 2003.

———. "Why the Sky Was Red in Munch's *Scream*." December 10, 2003.

Coates, Sam. "Friends Abroad Couldn't Save Shah from Enemies at Home." *The Times* (London), February 9, 2004.

Cook, Helen. "Sexed Down: Tennyson Altered His Charge of the Light Brigade to Make It Patriotic." *The Mirror*, January 30, 2004.

Cowley, Robert, ed. *What Ifs? Of American History*. New York: G. P. Putnam's Sons, 2003.

Craig, Leigh Ann. "Royalty, Virtue, and Adversity: The Cult of King Henry VI." *Albion*, June 22, 2003.

Creasy, Edward Shepherd. *Fifteen Decisive Battles of the World: From Marathon to Waterloo*. New York: A. L. Burt, 1851.

Creedy, Steve. "Beautiful Balloons." *The Australian*, December 24, 2004.

Crigger, Bette-Jane. "Energy Enters Guilty Plea." *Bulletin of the Atomic Scientists* (March 1994).

Crowd Dynamics, Ltd. http://www.crowddynamics.com.

Crumm, David. "Untold Stories of '67 Riot Divide Our Region, Our Lives." Knight Ridder, October, 19, 2004.

Current Events. "Vanished into Thin Air." September 6, 2002.

Curtin, Nancy, J. *The United Irishmen: Popular Politics in Ulster and Dublin, 1791–1798*. Oxford: Clarendon Press, 1994.

Daily Mail. "Atom Test Fallout Killed Thousands." March 1, 2002.

———. "Cloud Cover Sealed Fate of Nagasaki." March 31, 2001.

Daily Telegraph. "Obituary of Major-General Charles Sweeney; Pilot Who Flew the B-29 Superfortress Which Dropped the Atomic Bonb to Kill 70,000 People in Nagasaki." July 21, 2004.

D'Alto, Nick. "A Stroke of Genius: On Its 250th Anniversary, Benjamin Franklin's Famous Kite Experiment Reveals the Science Behind the Folklore." *Weatherwise* (May 2002).

Davis, William Stearns. *A History of France from the Earliest Times to the Treaty of Versailles*. Boston: Houghton Mifflin, 1919.

DeBlieu, Jan. *Wind: How the Flow of Air Has Shaped Life, Myth, and the Land*. New York: Houghton Mifflin, 1998.

DeGroot, Gerard. "The Afterlife of a Nuclear Test Site." *History Today* (June 2004).

Dennis, Jerry. *It's Raining Frogs and Fishes*. New York: Harper Perennial, 1992.

Devaney, Robert. "A Key Point in U.S. History." *Washington Times*, September 11, 1997.

Diamond, Jared. *Collapse: How Societies Choose to Fail or Succeed*. New York: Viking, 2005.

———. *Guns, Germs, and Steel*. New York: W. W. Norton, 1999.

Dickson, James H., et al. "The Iceman Reconsidered." *Scientific American* (February 2005).

Dill, Samuel. *Roman Society in Gaul in the Merovingian Age*. London: Macmillan, 1926.

Discover. "A Global Winter's Tale." (December 1998).

DISCovering World History. Farmington Hills, MI: Gale Research, 1997.

DISCovering World History. "Revolution Sentiments Spread Through Paris (personal account) June 9, 1789." Farmington Hills, MI: Gale Research, 1997. Reproduced in History Resource Center. http://www.galenet.galegroup.com/servlet/HistRC.

Dobson, Roger. "Feel Bad? Blame It on Weather." *Rocky Mountain News*, September 29, 1997.

Dopsch, Alfons. *The Economic and Social Foundations of European Civilization* London: K. Paul, 1937.

Doyle, William. "The Execution of Louis XVI and the End of the French Monarchy." *History Review* (March 2000).

Drake, David. "*Les temps modernes* and the French War in Indochina." *Journal of European Studies* 28 (March 1998).

Dreyer, Peter H. "Last Letters from Stalingrad." *World War II* (January 2003).

Drye, Willie. "America's Lost Colony: Can New Dig Solve Mystery?" *National Geographic News*, March 2, 2004.

Durschmied, Erik. *The Hinge Factor*. New York: Arcade Publishing, 2000.

———. *The Weather Factor*. New York: Arcade Publishing, 2001.

Earth Explorer. "Volcanoes and Weather." February 1, 1995.

Eaton, Mark. "War of the Roses' Second Round." *Military History*, December 1996.

Eberhart, Jonathan. "Another Launch Failure." *Science News*, April 4, 1987.

———. "*Challenger* Disaster: Rooted in History." *Science News*, June 14, 1986.

———. "Launch Score: Nature 3, NASA, 0." *Science News*, June 20, 1987.

Eden, Philip. "Cities That Can't Sleep." *The Daily Telegraph*, August 16, 2003.

Edgerton, Robert. *Death or Glory: The Legacy of the Crimean War*. Boulder, CO: Westview Press, 1999.

Egerton, Douglas R. *Gabriel's Rebellion: The Virginia Slave Conspiracies of 1800 and 1802*. Chapel Hill, NC: University of North Carolina Press, 1993.

Eisler, Peter. "Fallout Likely Caused 15,000 Deaths." *USA Today*, February 28, 2002.

Elegant, Robert. "Fallout: In Kazakhstan, the Human Wreckage of Soviet Nuclear Tests." *National Review*, September 16, 2002.

Elrick, M. L. "Mayor Urges Suburban Dwellers: Take a New Look at Detroit." *Detroit Free Press*, May 19, 2004.

Elton, Geoffrey. *The English*. Oxford: Blackwell Publishers, 1992.

Encyclopaedia Britannica, 2003 ed.

Encyclopedia of the Ancient Greek World. New York: Facts on File, 1995.

Encyclopedia of the Middle Ages. New York: Facts on File, 1995.

Encyclopedia of the Roman Empire. New York: Facts on File, 1994.

Encyclopedia of Ukraine. http://www.encyclopediaofukraine.com.

Encyclopedia of World Biography, 2nd ed. Detroit, MI: Gale Research, 1998.

Engle, Eloise, and Lauri Paananen. *The Winter War: The Russo-Finnish Conflict, 1939–1940*. New York: Charles Scribner's Sons, 1973.

Erdoes, Richard. *A.D. 1000: A World on the Brink of Apocalypse*. San Francisco, CA: Harper and Row, 1988.

Erickson, John. "Nazi Posters in Wartime Russia." *History Today* (September 1994).

Events of the Middle Ages. New York: Great Neck Publishing, 2002.

Fagan, Brian. *Floods, Famines, and Emperors: El Niño and the Fate of Civilization*. New York: Basic Books, 1999.

———. *The Little Ice Age: How Climate Made History 1300–1850*. New York: Basic Books, 2000.

———. *The Long Summer*. New York: Basic Books, 2004.

Fenton, Ben. "Invasion Weather Forecasters Sailed Close to the Wind." *The Daily Telegraph*, June 5, 2004.

Fernandez-Armesto, Felipe. "What If the Armada Had Landed . . ." *New Statesman*, December 20, 1999.

Findling, John, and Frank W. Thackeray, eds. *Events That Changed the World Through the Sixteenth Century*. Westport, CT: Greenwood Press, 2001.

Fine, Sidney. *Violence in the Model City*. Ann Arbor, MI: University of Michigan Press, 1989.

Flanagan, Laurence. "Wrecks of the Spanish Armada." *Natural History* (September 1988).

Flatow, Ira. *They All Laughed*. New York: HarperCollins, 1992.

Flynn, Sian. "The Race to the South Pole." BBC, August 1, 2002. http://www.bbc.co.uk/history.

Folkers, Richard. "Everyday Mysteries." *U.S. News and World Report*, August 18, 1997.

Fong, Chua Lu. "Operation Eagle Claw, 1980: A Case Study in Crisis Management and Military Planning." *Journal of the Singapore Armed Forces* 28, no. 2 (April–June, 2002).

Fordney, Chris. "The Long Road to Andersonville." *National Parks* (September 1998).

Foss, Clive. "Russia's Romance with the Airship." *History Today* (December 1997).

Fowler, Christopher. "When the Sky Went Dark Over Chelsea." *The Independent on Sunday*, May 25, 2003.

Freehoff, William Francis. "Tecumseh's Last Stand." *Military History* (October 1998).

Friedrich, Otto. "Every Man Was a Hero." *Time*, May 28, 1984.

Gani, Martin. "A Cremonese Sound—The Violin Makers of Cremona, Italy." *World and I* (February 2001).

Garrison, Webb. *Lost Pages from American History*. Harrisburg, PA: Stackpole Books, 1976.

Gavrilovich, Peter, and Bill McGraw, eds. *The Detroit Almanac: 300 Years of Life in the Motor City.* Detroit, MI: Detroit Free Press, 2000.

Gelbert, Doug. *So Who the Heck Was Oscar Mayer?* New York: Barricade Books, 1996.

Gessen, Masha. "A Tale of One City." *U.S. News and World Report*, February 17, 2003.

Given-Wilson, Chris. *The English Nobility in the Late Middle Ages: The Fourteenth-Century Political Community.* London: Routledge, 1996.

Glantz, David M., and Jonathan House. *When Titans Clashed: How the Red Army Stopped Hitler*. Lawrence, Kans: University Press of Kansas, 1995.

Goulding, Vincent J., Jr. "Back to the Future with Asymmetric Warfare." *Parameters* 30, no. 4 (2000).

Graham, Mark. "Providence Flows." *Weatherwise* (July 2002).

Gray, Paul. "Doomsdays." *Time* 146, no. 6 (August 7, 1995).

Greely, M. Sgt. Jim. "Desert One." *Airman*, April 1, 2001.

Griffin, Sally. "Blast from the Past: The Gulf War Syndrome U-Turn Should Remind Us of the Uncompensated Victims of Our H-Bomb Tests." *New Statesman*, May 16, 1997.

Guttman, Jon. "The Irish Won a Great Victory Four Centuries Ago." *Military History* (August 1998).

Gwybodiadur, a Welsh Informationary. http://www.gwybodiadur.co.uk.

Haigh, Philip. *The Military Campaigns of the Wars of the Roses*. Conshocken, PA: Combined Books, 1997.

Hamelink, Jerry H. "Instrumentation and the Art of Violin Making." *Mechanical Engineering* (October 1989).

Hanlon, Michael. "Supervolcano!" *Daily Mail*, December 3, 2004.

Hansen, Chuck. "The Oops List." *Bulletin of the Atomic Scientists* 56, no. 6, (2000).

Hastings, Max, ed. *The Oxford Book of Military Anecdotes*. New York: Oxford University Press, 1985.

Hatcher, Patrick Lloyd. *North Atlantic Civilization at War: The World War II*

Battles of Sky, Sand, Snow, Sea, and Shore. Armonk, NY: East Gate Books, 1998.

Hawkins, William R. "The Bomb Is Dropped on Hiroshima." Knight Ridder, July 28, 1995.

Hawks, Francis. *The History of North Carolina*, vol. 1. Fayetteville, NC: E. J. Hale and Son, 1857.

Hecht, Jeff. "Black Sea Bore the Brunt of Two Gushing Neighbours." *New Scientist*, July 26, 2003.

Henderson, Mark. "Eruption That Could Wipe Out Millions." *The Times* (London), March 9, 2005.

History Today. "The Irish Rising of 1798." June 1998.

Holzman, Benjamin G. "The Effects of Atomic Bombs on the Weather." *Weatherwise* (January 1998).

Holzworth, C. E. "Operation Eagle Claw: A Catalyst for Change in the American Military." http://www.globalsecurity.org/military/library/report/1997/Holzworth.htm.

Hopkinson, Deborah. "The Volcano That Shook the World: Krakatoa 1883." *Storyworks* 11, no. 4 (2004).

Hornblower, Margot. "Liberté, Egalité, Fraternité?" *Time*, May 1, 1989.

Hudson, Christopher. "The Other Holocaust." *Daily Mail*, June 23, 2001.

Hughes, Patrick. "Winning the War." *Weatherwise* (June 1995).

Inbaraj, Sonny. "Australia: Birth Defects from British Nuclear Tests." Inter Press News Service English News Wire, January 12, 1998.

Iran Chamber Society. http://www.iranchamber.com.

Ives, John. "Books: Within a Whisker of Defeat?" *Birmingham Post*, June 5, 2004.

Jackson, Carlton. "The Kitchener Coffin Caper." *British Heritage* (August/September, 1997).

James, Barry. "Airships Poised to Float Back into View." *International Herald Tribune*, July 24, 2000.

Jamieson, J. W. "The Battle That Stopped Rome." *Mankind Quarterly* (Winter 2003).

Jankovic, Vladimir. *Reading the Skies: A Cultural History of English Weather, 1650–1820*. Chicago: University of Chicago Press, 2000.

Johnson, Rob. "Turning Up the Heat." *Shopping Center World* (September 2000).

Jones, Jerome. "Krakatau: Sundra Straits Indonesia." http://www.brookes.ac.uk/geology/8361/1997/jerome.html.

Kamata, Sadao. "The Atomic Bomb and the Citizens of Nagasaki." *Bulletin of Concerned Asian Scholars* 12, no. 2 (1982).

Kansas City Star. "How Literally Should We Take the Story of Noah and the Flood?" December 3, 2003.

Kauffman, Bill. "O Say Can You Sing?" *American Enterprise* (October–November 2003).

Keegan, John. "Bad Weather Could Have Scuppered D-Day." *The Daily Telegraph*, June 7, 2004.

Kendall, Paul Murray. *The Yorkist Age: Daily Life During the Wars of the Roses*. New York: Norton, 1962.

Kennedy, Bruce. "Nuclear Close Calls." http://www.cnn.com/Specials/cold.war.

Keys, David. *Catastrophe: An Investigation into the Origins of the Modern World*. New York: Ballantine, 1999.

Khariton, Yuli, and Yuri Smirnov. "The Khariton Version: Only a Handful of Men Knew the Full Story of the Soviet Bomb's Creation." *Bulletin of the Atomic Scientists* 49, no. 4 (May 1, 1993).

Kidder, Chris. "Museum Spread Some Wrong Information." *Virginian Pilot*, November 2, 2003.

———. "Wright Brothers Were Little Known in Flying Community." *Virginian Pilot*, November 18, 2001.

Kingwell, Jeff, et al. "Weather Factors Affecting Rocket Operations: A Review and Case History." *Bulletin of the American Meteorological Society* (June 1991).

Kitchener, Horatio. "Kitchener's Address to the Troops." *Essential Documents in American History, Essential Documents, 1492–Present*. Academic Search Premier Database. (*AN 9708140293*).

Kite, Carly. "Weathering History: From Cold War Espionage to Space-Age Triumph. Retired Air Force Meteorologist Lieutenant Colonel Hank Brandli Remembers Forecasting for Critical Moments in American History." *Weatherwise* (May 2003).

Klingspor, Carl Gustafson. *Charles the Twelfth, King of Sweden*. Translated by John A. Gade. Boston: Houghton Mifflin, 1916.

Konstam, Angus. *Historical Atlas of the Crusades*. New York: Checkmark Books, 2002.

Landrum, Armetta. "Does Summertime Heat Spark Violence?" *Call and Post* (Cleveland), June 22, 1995.

Latimer, Jon. "French Farce at Fishguard." *Military History* (March 1997).

———. "Storm of Snow and Steel." *Military History* (December 2000).

Laurence, William L. "Atomic Bombing of Nagasaki Told by Flight Member," *New York Times*, September 9, 1945.

Lawder, David. "Detroit Erasing Scars of 1967 Riot." *Denver Rocky Mountain News*, August 10, 1997.

Lazar, Barry. "Eating Seal." http://www.montrealfood.com/eatingseal.html.

Lederer, Richard. *The Miracle of Language*. New York: Pocket Books, 1991.

Lee, Albert. *Weather Wisdom*. New York: Doubleday, 1978.

Lee, Joe. "The Road to Wexford." *World of Hibernia*, March 22, 1998.

Lee, Laura. *Bad Predictions*. Rochester, MI: Elsewhere Press, 2000.

———. *100 Most Dangerous Things in Everyday Life and What You Can Do About Them*. New York: Broadway Books, 2004.

———. *The Name's Familiar: Mr. Leotard, Barbie, and Chef Boyardee*. Gretna, LA: Pelican Publishing, 1999.

———. *The Name's Familiar II*. Gretna, LA: Pelican Publishing, 2001.

Legassé, Paul. *Columbia Encyclopedia*, 6th ed. New York: Columbia University Press, 2000.

Lekic, Slobodan. "Adventure Seekers Drawn to New Krakatau Volcano." *Cincinnati Post*, June 18, 2004.

———. *The Mammoth Book of Journalism*. New York: Carroll and Graf, 2003.

Lemonick, Michael D., and Andrea Dorfman, "The Amazing Vikings." *Time* 155, no. 19 (May 8, 2000).

Levack, Brian. "Witch Trials." http://www.utexas.edu/features/archive/2004/witches.html.

LeVine, Steve. "A Half Century of Nuclear Blasts: And the Environmental Fallout Is Just Beginning." *Newsweek International*, September 13, 1999.

Lewis, Jon E., ed. *The Mammoth Book of Eyewitness World War II*. New York: Carroll and Graf, 2002.

Lichfield, John. "Agincourt Remembers Battle Lost in Mists of Time." *The Independent*, August 11, 2001.

———. "Waterloo's Significance to the French and British." *The Independent*, November 17, 2004.

Lindaman, Dana, and Kyle Ward. *History Lessons: How Textbooks from Around the World Portray U.S. History*. New York: The New Press, 2004.

Lindgren, S., and J. Neumann. "Great Historical Events That Were Significantly Affected by the Weather: 5, Some Meteorological Events of the Crimean War and Their Consequences." *Bulletin of the American Meteorological Society* 61, no. 12 (1980).

———. "Great Historical Events That Were Significantly Affected by the Weather: 6, Inundations and the Mild Winter of 1672–73 Help Protect Amsterdam from French Conquest." *Bulletin of the American Meteorological Society* 64, no. 7 (1983).

———. "Great Historical Events That Were Significantly Affected by the Weather: 7, 'Protestant Wind'–'Popish Wind': The Revolution of 1688 in England." *Bulletin of the American Meteorological Society* 66, no. 6 (1985).

Loewen, James W. *Lies My Teacher Told Me*. New York: The New Press, 1995.
Long, Phil. "Lethal Plutonium on Rocket Triggers Alarm." Knight Ridder, August 23, 1997.
Ludlu, David. *The Weather Factor*. Boston: Houghton Mifflin, 1984.
Lusbrink, Hans-Jurgen, et al. *The Bastille: A History of a Symbol of Despotism and Freedom*. Durham, NC: Duke University Press, 1997.
Lyons, Walter. *The Handy Weather Answer Book*. Detroit, MI: Visible Ink Press, 1997.
Maddock, Robert K. "The Finnish Winter War." http://www.wargamesdirectory.com.
Magnuson, Ed. "Last Flight of *Challenger*'s Crew." *Time*, May 12, 1986.
Maier, Timothy W. "Weather Warfare." *Insight on the News*, April 21, 1997.
Makhijani, Arjun. "Always the Target: While U.S. Bomb Scientists Were Racing Against Germany, Military Planners Were Looking Toward the Pacific." *Bulletin of the Atomic Scientists* (May 1995).
———. "Worse Than We Knew." *Bulletin of the Atomic Scientists* (November 1997).
Matthews, Robert. "Benjamin Franklin 'Faked Kite Experiment' New Book Says." *The Sunday Telegraph*, June 1, 2003.
McAdie, Alexander. *War Weather Vignettes*. New York: Macmillan, 1925.
McCrum, Robert, et al. *The Story of English*. New York: Viking, 1986.
McNeill, William H. *Plagues and People*. Magnolia, MA: Peter Smith Publisher, 1992.
Medieval Sourcebook. http://www.fordham.edu/halsall/sbook.html.
Melikian, Souren. "The Franks: Their Times and Treasures." *International Herald Tribune*, June 14, 1997.
Mercer, Patrick. "The Complete Guide to Crimea." *The Independent*, March 27, 2004.
Merrett, J. "Burning with Rage." *The Mirror*, June 1, 2001.
Meteorology Education and Training. http://www.meted.ucar.edu/.
Methvin, Eugene H. *The Rise of Radicalism: The Social Psychology of Messianic Extremism*. New Rochelle, NY: Arlington House, 1973.
Michel, Dieter. "Villains, Victims, and Heroes: Contested Memory and the British Nuclear Tests in Australia." *Journal of Australian Studies* (January 2004).
Michigan Citizen. "30th Anniversary: 1967 Detroit Rebellion." July 16, 1997.
Milman, Henry Hart. "Election of Antipope Clement VII: Beginning of the Great Schism." *History of the World* (January 1992).
Mitchell, Joseph B., and Edward Creasy. *Twenty Decisive Battles of the World*. Old Saybrook, CT: Konecky and Konecky, 1964.

Mollat, Guillaume. *The Popes at Avignon, 1305–1387*. Translated by Janet Love. New York: T. Nelson and Sons, 1963.

Monastersky, R. "Meteorologists Follow Franklin's High-Flying Legacy." *Science News*, April 4, 1992.

Montgomery, Christine. "Collectors Report Rush for Old News." *Washington Times*, April 29, 1998.

Moore, Richard. "Where Her Majesty's Weapons Were." *Bulletin of the Atomic Scientists* (January 2001).

Moore, Victoria. "Enough to Make You Scream." *Daily Mail*, August 24, 2004.

Morris, Charles. *The Marvelous Record of the Closing Century*. Philadelphia, PA: American Book and Bible House, 1899.

Morrison, Herb. "Crash of the *Hindenburg*." *Essential Speeches*, 2003.

Morse, W. E. *The Rise and Fall of the Crimean System, 1855–71: The Story of a Peace Settlement*. London: Macmillan, 1963.

Murphy, Jamie. "Dateline: Aboard the Shuttle; the Nation's Journalists Compete to Cover a Far-Out Story." *Time*, January 27, 1986.

Neumann, J. "Great Historical Events That Were Significantly Affected by the Weather: 1. The Mongol Invasions of Japan." *Bulletin of the American Meteorological Society* (November 1975).

———. "Great Historical Events That Were Significantly Affected by the Weather: 2. The Year Leading to the Revolution of 1789 in France." *Bulletin of the American Meteorological Society* (February 1977).

———. "Great Historical Events That Were Significantly Affected by the Weather. Part 11: Meteorological Aspects of the Battle of Waterloo." *Bulletin of the American Meteorological Society* (March 1993).

———. "The Sea and Land Breezes in the Classical Greek Literature." *Bulletin of the American Meteorological Society* (January 1973).

Neumann J., and J. Dettwiller. "Great Historical Events That Were Significantly Affected by the Weather: 9, The Year Leading to the Revolution of 1789 in France (II)." *Bulletin of the American Meteorological Society* (January 1990).

Neumann J., and H. Flohn. "Great Historical Events That Were Significantly Affected by the Weather: 8, Germany's War on the Soviet Union, 1941–45. Long-Range Weather Forecasts for 1941–42 and Climatological Studies." *Bulletin of the American Meteorological Society* (June 1987).

———. "Great Historical Events That Were Significantly Affected by the Weather: 8, Germany's War on the Soviet Union, 1941–45. II. Some Important Weather Forecasts, 1942–45." *Bulletin of the American Meteorological Society* (July 1988).

New Internationalist. "A Brief History of Slavery." August 1, 2001.

Nobile, Philip. "Hiroshima, Nagasaki After Atomic Blasts: Story the Smithsonian Was Not Allowed to Tell." *National Catholic Reporter* 31, no. 35 (July 1995).

O'Brien, David. "O Lucky Canada." *Bulletin of the Atomic Scientists* (July 2003).

Oman, Charles W. *Warwick, The Kingmaker*. London: Macmillan, 1903.

Oppenheimer, Stephen. "The First Exodus." *Geographical* (July 2002).

Ormrod, W. M. "Edward III: W. M. Ormrod Describes the Career of the King Whose Fifty Years on the Throne Are Best Remembered for His Wars with France and Scotland." *History Today* (June 2002).

Oster, Emily. "Witchcraft, Weather, and Economic Growth in Renaissance Europe." *Journal of Economic Perspectives* (Winter 2004).

Patterson, Margot. "Papal Echoes in Avignon." *National Catholic Reporter*, April 16, 2004.

Paul, Annie Murphy. "Hot and Bothered." *Psychology Today* (May/June 1998).

Perkins, Allison. "Witness to Destruction: Servicemen Recall Visiting Hiroshima Weeks after Atomic Bomb Dropped." *News and Record* (Piedmont Triad, NC), August 6, 2002.

Perkins, Sid. "Full Report of Nuclear Fallout Test Released." *Science News*, October 11, 1997.

Perrow, Charles. "Risky Systems; The Habit of Courting Disaster." *The Nation*, October 11, 1986.

Pershing, Ben. "Stormy Tuesday?" *Roll Call*, November 4, 2002.

Peters, Charles. "From Ouagadougou to Cape Canaveral: Why the Bad News Doesn't Travel Up." *Washington Monthly*, April 1986.

Pickerell, John. "Did Little Ice Age Create Stradivarius Violins' Famous Tone?" *National Geographic News*, January 7, 2004.

Pope, Eric. "Fallen City Looks to Rise." *Crain's Detroit Business*, July 2, 2001.

Poroskov, Nikolai. "Stalingrad: The Battle That Broke Hitler's Back." *Russian Life* (November–December 2002).

Prest, Wilfrid. *Albion Ascendant: English History, 1660–1815*. Oxford: Oxford University Press, 1998.

Preston, John. "The Island That Went Bang." *Sunday Telegraph*, June 1, 2003.

Pry, Peter Vincent. *War Scare: Russia and America on the Nuclear Brink*. Westport, CT: Praeger, 1999.

Purcell, L. Edward, and Sarah J. Purcell. *Encyclopedia of Battles in North America: 1517 to 1916*. New York: Facts on File, 2000.

Quinion, Michael. "World Wide Words." http://www.worldwidewords.org.

Rainie, Harrison, et al. "Requiem for the Cities?" *U.S. News and World Report*, May 18, 1992.

Rampino, Michael R., and Stephen Self. "Volcanic Winter and Accelerated Glaciation Following the Toba Super-eruption." *Nature*, September 3, 1992.

Ramsayer, K. "Frosty Florida: Spread of Agriculture May Promote Freezes." *Science News*, November 8, 2003.

Rather, Dan, and Byron Pitts. "NASA Scientists Study Urban 'Heat Island.'" CBS Evening News, June 11, 1999.

Ratnasabapathy, Senthil. "Environment: Ex-Soviet Nuclear Activity Raises Radiation Exposure." Inter Press Service English News Wire, May 23, 1995.

Reader's Digest Association. *The Truth About History*. Pleasantville, NY: *Reader's Digest*, 2003.

Reed, Susan E. "Atomic Lake: The Victims of Soviet Nuclear Testing." *New Republic*, October 28, 1991.

Rees, Nigel. *The Cassell Dictionary of Anecdotes*. London: Cassell, 1999.

Regan, Geoffrey. *The Past Times Book of Military Blunders*. London: Guinness Publishing, 1991.

Resnik, Abraham. *Due to the Weather: Ways the Elements Affect Our Lives*. Westport, CT: Greenwood Press, 2000.

Richardson, Sarah. "Vanished Vikings: Erik the Red's Greenland Colonies." *Discover* (March 2000).

Richman, Alan. "A Lesson in Uncommon Valor." *People Weekly*, February 10, 1986.

Ridley, Jane. "55 Years on from Hiroshima: The Dark Nuclear Clouds Still Hanging Over Us All." *Mirror*, August, 7, 2000.

Riley-Smith, Jonathan. "Rethinking the Crusades." *First Things: A Monthly Journal of Religion and Public Life* (March 2000).

Riley-Smith, Jonathan, ed. *The Oxford History of the Crusades*. Oxford: Oxford University Press, 1999.

Ring, Natalie. "Inventing the Tropical South: Race, Region, and the Colonial Model." *The Mississippi Quarterly* 56, no. 4 (2003).

Rivera, Rachel. "Hot Spot: Scientists Probe Beneath Yellowstone and Find a Seething Volcanic Basin That Could One Day Blow Its Top." *Science World*, February 25, 2002.

Roberts, Geoffrey. "Stalingrad." *History Review* (December 2004).

Roberts, Royston. *Serendipity: Accidental Discoveries in Science*. New York: John Wiley and Sons, 1989.

Robinson, J. H., ed. *Readings in European History*. Boston: Gin and Company, 1904.

Room, Adrian. *Brewer's Names: People, Places, Things*. London: Cassell, 1992.

Rosenblum, Mort. "French Colonel's War Diary Recounts 56 Days at Dien Bien Phu." AP Worldstream, April 29, 2004.

Rosenfeld, Jeffrey. *Eye of the Storm*. New York: Plenum Press, 1999.

Ryan, William, and Walter Pitman. *Noah's Flood*. New York: Simon and Schuster, 1998.

Salem Press. *Great Scientific Achievements*. Pasadena, CA, 1994.

Schwartz, Stephen I. "Atomic Audit: What the U.S. Nuclear Arsenal Has Cost." *Brookings Review*, September 22, 1995.

Science News. "Exploding into an Ice Age." September 26, 1992.

———. "Icy Tales of Ancient Eruptions." October 15, 1994.

———. "When Tempers and Temperatures Flare." March 3, 1984.

Scott, Gale. "Flash Back." *Weatherwise* (July/August 2003).

Seabrook, Charles. "Atlanta's Heat Believed to Spawn Storms." *Atlanta Constitution*, May 4, 2000.

Seattle Post-Intelligencer. "Idaho Falls Sees Rise in Thyroid Cancer Rate." August 18, 1997.

———. "Rural and Central Idaho Gets the Radiation News Way Late." September 8, 1997.

Shafritz, Jay M. *Words on War: Military Quotations from Ancient Times to the Present*. New York: Simon and Schuster, 1990.

Sheler, Jeffrey L. "Say a Little Prayer for You." *U.S. News and World Report*, July 8, 2002.

Simonds, Frank. *History of the World War*. Garden City, NY: Doubleday, 1917.

Simons, Paul. "Crimean War Put Forecasts on the Map." *The Times* (London), November 13, 2004.

———. "Weather Eye." *The Times* (London), June 17, 2002.

———. "Weather Eye." *The Times* (London), August 15, 2002.

———. "Weather Secrets of Miracle at Fátima." *The Times* (London), February 17, 2005.

Simons, Paul, and Robin Young. "The Day Weathermen Made the News." *The Times* (London), June 4, 2004.

Smith, Walter Bedell. *Eisenhower's Six Great Decisions: Europe, 1944–1945*. New York: Longmans, Green, 1956.

Soroos, Marvin S. "Preserving the Atmosphere in the Global Commons." *Environment* 40, no. 2 (1998).

Spence, Clark C. *The Rainmakers: American "Pluviculture" to World War II*. Lincoln, Neb: University of Nebraska Press, 1980.

Spotts, Peter, N. "Black Sea Find May Explain Noah's Flood." *Christian Science Monitor*, September, 15, 2000.

Stead, Mark. "Ill-Fated Expedition Made Scott a Hero." *South Wales Echo*, June 15, 2004.

Steinberg, Theodore. *Acts of God: The Unnatural History of Natural Disaster in America*. New York: Oxford University Press, 2000.

———. *Down to Earth: Nature's Role in American History*. New York: Oxford University Press, 2002.

Steinmetz, Leon. "The Bolsheviks of the Bastille." *National Review*, July 14, 1989.

St. Louis Post-Dispatch. "Weather May Make Space Teacher Tardy." January 26, 1986.

Stuller, Jay. "It's a New Battle Every Day in the War on Whiskers." *Smithsonian* (February 1995).

Svitil, Kathy A. "The Greenland Viking Mystery." *Discover* (July 1997).

Tanikawa, Miki. "Tokyoites Feel the Heat." *International Herald Tribune*, August 28, 2002.

Tarm, Michael. "This Europe: Mass Grave Shows How Hunger and Cold Devastated Napoleon's Army on Retreat from Moscow." *The Independent*, September 3, 2002.

Taylor, Frank, and John S. Roskell, trans. *Gesta Henrici Quinti: The Deeds of Henry the Fifth*. Oxford: Clarendon Press, 1975.

The Times (London). "The Bad: Death, Violence, Smog and Vicious Bugs." August 9, 2003.

The Times (London). "Weather Eye." June 15, 2004.

Thomas, Pat. *Under the Weather: How the Weather and Climate Affect Our Health*. London: Fusion Press, 2004.

Time. "Wandering into Trouble (John Weymouth Wanders Across Bering Strait)." April 28, 1986.

Time International. "Under a Cold War Cloud." March 30, 1998.

The Times (London). "On This Day, May 7, 1937." May 7, 2004.

Trefil, James. "Evidence for a Flood." *Smithsonian* (April 2000).

Trulock, Alice Rains. *In the Hands of Providence: Joshua L. Chamberlain and the American Civil War*. Chapel Hill, NC: University of North Carolina Press, 1992.

Turner, Steve. *Amazing Grace: The Story of America's Most Beloved Song*. New York: HarperCollins, 2002.

Udall, Stewart L. *The Myths of August: A Personal Exploration of Our Tragic Cold War Affair with the Atom*. New York: Pantheon, 1994.

UPI. "*Challenger* Accident Still Impacts NASA." January 25, 2001.

Van Biema, David H. "Christa McAuliffc Gets NASA's Nod to Conduct America's First Classroom in Space." *People Weekly*, August 5, 1985.

Van Ells, Mark D. "A Taste for War: A Culinary History of the Blue and the Gray." *Journal of Popular Culture* (November 2004).

Vincent, John. "Hiroshima Pilot Details Fateful Flight." *Evening Standard*, March 25, 2002.

Volti, Rudi. *The Facts on File Encyclopedia of Science, Technology, and Society*, vol. 1. New York: Facts on File, 1999.

Waldman, Carl. *Atlas of the North American Indian*. New York: Facts on File, 2000.

Walvin, James. *Black Ivory: A History of British Slavery*. Washington, DC: Howard University Press, 1994.

Warner, Gary A. "Civil War Prison Also Home to National Prisoner of War Museum." *Orange County Register*, June 18, 2003.

Washington Times. "Andersonville Chief Viewed Less Critically." November 15, 2003.

Weatherwise. "Way Up High." September 2004.

Weaver, Kenneth F. "The Search for Our Ancestors: Stones, Bones, and Early Man." *National Geographic* (November 1985).

Western Mail. "US's Darkest Day as 31 Die in Helicopter Crash." January 27, 2005.

Wheeler, Linda. "Civil War Prison Made Headlines." *Washington Post*, June 17, 2004.

White, John. "The Man Behind Amazing Grace." *The New American*, August 9, 2004.

White, Thomas G. "The Establishment of Blame in the Aftermath of a Technological Disaster." *National Forum* (Winter 2001).

Wills, Christopher. *Yellow Fever, Black Goddess: The Coevolution of People and Plagues*. Boston: Addison Wesley, 1996.

Wilson, David A. *The United Irishmen, United States: Immigrant Radicals in the Early Republic*. Ithaca, NY: Cornell University Press, 1998.

Wilson, Jim. "Weather Wars." *Popular Mechanics*, February 1, 1997.

Winters, Harold A. *Battling the Elements: Weather and Terrain in the Conduct of War*. Baltimore: Johns Hopkins University Press, 1998.

Witze, Alexandra. "Scientist Links Famous 'Scream' Painting to Devastating Volcano." *Dallas Morning News*, December 9, 2003.

Wood, Ian. *The Merovingian Kingdoms: 450–751*. London: Longman, 1994.

Wood, James. "A Voice of Reason in the Midst of the European Witch Hunts." *Skeptical Inquirer* (March 2005).

Wood, Richard. "Hail: The White Plague." *Weatherwise* (April 1993).

Wright, John D. *The Language of the Civil War*. Westport, CT: Oryx Press, 2001.

Wright, Mike. *What They Didn't Teach You About the Civil War*. Novato, CA: Presidio Press, 1996.

Wuorinen, John H, ed. *Finland and World War II, 1939–1944*. New York: Ronald Press, 1948.

Xinhua News Agency. "Vietnam Marks 50th Anniversary of Dien Bien Phu Victory." May 5, 2004.

Youngson, R. M. *Scientific Blunders: A Brief History of How Wrong Scientists Can Sometimes Be*. New York: Carroll and Graf, 1998.

Zamoyski, Adam. *Moscow 1812: Napoleon's Fatal March*. London: HarperCollins, 2004.

Zimmermann, Tim. "Just When You Thought You Were Safe." *U.S. News and World Report*, February 10, 1997.

IF YOU LIKED *BLAME IT ON THE RAIN*, YOU'LL LOVE . . .

WHY GIRLS CAN'T THROW

And Other Questions You Always Wanted Answered

ISBN 0-06-083518-4 (paperback)

Symons attempts to answer the often pondered but never explained questions of our lives. He interviews experts, researches on the Internet, and looks to the wisdom of friends to satisfy his ever curious brain. From the moral—Is it wrong to burn a CD before giving it to a friend as a gift?—to the practical—Are you better off walking or running through the rain?—Symons never backs away from his task of answering the eclectic questions of life.

THAT BOOK

. . . of Perfectly Useless Information

ISBN 0-06-073149-4 (hardcover)

That Book is quite unlike anything you have ever read. Frighteningly addictive and almost entirely useless information is packed onto every page—once you dip in you'll be hooked for hours.

THIS BOOK

. . . of More Perfectly Useless Information

ISBN 0-06-082823-4 (hardcover)

This Book has everything, from tongue twisters, to things invented by women, to the working titles of Beatles songs. There are celebrities who own islands, went bankrupt, had liposuction, survived helicopter crashes, had hotel suites named after them, married (and divorced) their childhood sweethearts, and more.

Visit www.AuthorTracker.com for exclusive information on your favorite HarperCollins authors.

Available wherever books are sold, or call 1-800-331-3761 to order.